ASK THE BEASTS:
DARWIN AND THE GOD OF LOVE

ASK THE BEASTS:

Darwin and the God of Love

Elizabeth A. Johnson

BLOOMSBURY

LONDON · OXFORD · NEW YORK · NEW DELHI · SYDNEY

Bloomsbury Continuum
An imprint of Bloomsbury Publishing Plc

50 Bedford Square
London
WC1B 3DP
UK

1385 Broadway
New York
NY 10018
USA

www.bloomsbury.com

First published in Great Britain 2014

Paperback edition 2015

British Library Cataloguing-in-Publication Data
A catalogue record for this book is available from the British Library.

Library of Congress Cataloguing-in-Publication data has been applied for.

ISBN: HB: 9781472903730
PB: 9781472924018
ePDF: 9781472903754
ePub: 9781472903747

2 4 6 8 10 9 7 5 3 1

Typeset by Fakenham Prepress Solutions, Fakenham, Norfolk NR21 8NN
Printed and bound in Great Britain by CPI Group (UK) Ltd, Croydon CR0 4YY

I dedicate this book to
my religious order,
the Sisters of St. Joseph, Brentwood, New York;
to the wider Federation of the Sisters of St. Joseph;
and to the Leadership Conference of Women Religious;
faithful women on the journey,
it is an honor to walk in your company

CONTENTS

ACKNOWLEDGMENTS

This book had its genesis in a Fordham University faculty seminar which spent the 150th anniversary year of Darwin's *On the Origin of Species* (1859) slowly reading and discussing this classic text. Insights from my colleagues in biology, history, economics, literature, philosophy, and political science were riveting; the conversation prompted me to keep a running list of theological questions which became the core of the present book. My first thanks, therefore, go to the members of this seminar; to Michael Latham, Dean of Fordham College, who provided the funding for our books and suppers; and to Angela O'Donnell, Associate Director of the Curran Center for American Catholic Studies, who supplemented these funds and superbly organized our sessions. I so appreciate being a member of this vibrant university community.

Most of the preliminary work was done during a faculty research leave granted in the regular rotation, for which I am thankful. Beyond that, I owe an uncommon debt of gratitude to Stephen Freedman, Provost of Fordham University and himself a biologist. When certain public events took away the sabbatical time and energy intended for writing this book, he not only inquired about the state of my research but, saying that one of his main responsibilities was to protect and promote the work of the faculty, took practical steps to provide needed time and support. The concrete, imaginative, generous way this academic leader found to assist my work is extraordinary, and I offer him a thousand warm thanks. My appreciation also goes to the chair of my department, Terrence Tilley, who offered extra research assistance to help move the project along and to my supremely capable research assistants Kathryn Reinhard, Daniel Rober, Christine McCarthy, Brianne Jacobs, and Elizabeth Pyne, whose work on texts, library searches, and bibliography was invaluable.

While this book was underway, invitations to lecture allowed its ideas to be tested in public settings. The Templeton Foundation symposium "Is God Incarnate in All That Is?" gathered an international community of scholars in Denmark for discourse and debate. My great thanks to Mary Ann Meyers who shepherded the program and Niels Gregersen who hosted and chaired, both in a most gracious way. Further exchanges with discerning audiences in settings of higher education in the U.S. and Canada kept the thesis growing. These institutions included Duquesne University, University of California San Diego, Dominican University, Wake Forest University, King's University College, Molloy College, Fairfield University, and Boston College. Key arguments were honed in helpful conversation with Denis Edwards, himself writing on the same subject. My bright colleagues Mary Catherine Hilkert, Roger Haight, and Frank Oveis gave particular chapters a thoroughly critical reading. The whole manuscript was read and discussed by the Presidential Seminar on the Catholic Intellectual Tradition at Fairfield University, chaired by Paul Lakeland; their questions and robust feedback served to clarify some of the book's main arguments. To all who entered into conversation which sharpened and deepened this book's ideas, I am deeply grateful.

Robin Baird-Smith, publisher at Bloomsbury/Continuum, has been kind enough to serve as my editor. It has been a delight. He is a lovely person to work with and I salute him with sincerest thanks.

Heartfelt appreciation goes to my family and friends who have given constant, unstinting support as this project took shape.

My profound gratitude goes to the numerous readers near and far who by words and deeds have offered ongoing encouragement for my writing over these past years. This book is borne on the shoulders of their good will.

I dedicate this book to nuns, otherwise known as women religious: to my own religious community, first of all, whose leaders years ago charged me to study theology and whose members have been an unfailing strength and a shelter in storms; also to the federated network of St. Joseph Sisters around the world; and to the LCWR, the national group of leaders of women's religious orders in the U.S., who remain stalwart in showing grace under pressure. As I write I am nearing the 54th anniversary of my own entrance into religious life. The sweep of changes in church and world has been staggering. But the courageous fidelity and love shown by these women as individual Sisters, as communities, and as leaders freely committed to living the gospel in our day continues to be an unfailing beacon. Thank you.

PREFACE

In an artful essay on the history of the universe, Holmes Rolston proposes that there have been not one but three Big Bangs. Big Bang is used here as a metaphor for a singular, explosive event with radical consequences for generating novelty.[1] The first Big Bang was the primordial hot explosion that started the universe approximately 13.7 billion years ago. Generating an expanding magnitude of matter-energy, it initiated the process that has produced stars, galaxies, and everything else in the cosmos, a dynamic phenomenon that is still going on. Ten billion years later, using materials produced in the first Big Bang, the second one hatched life on Earth. Here began the evolutionary process that now covers our planet with beautiful, complex creatures interacting in life-sustaining ecosystems. More than three billion years later still, a third singular event shaped up on the shoulders of the first and second. This is the emergence of human beings, *Homo sapiens*, mammals with minds and wills who think symbolically and act with deliberate, free intent. There is no question but that many other living creatures experience emotion, enjoy sophisticated levels of knowing and communicating, and act with a certain purpose. In this regard humans belong on a spectrum with others in the community of life. The appearance of the human species is rightly considered a third moment of intense novelty, however, because of the qualitative prowess of the species as a whole. With extraordinary ability we (I write as one such mammal) have populated the globe with a restless inventiveness that creates, accumulates, and transmits ideas and technologies across generations. Evolution now proceeds by way of culture as well as biology. From matter to life to mind; from physical matter to biological life to linguistic consciousness; from galaxies to living species to human persons: though connected, these explosions form no simple, predictable unfolding but a fascinating, unexpected story.

This book pours out attention on the second big bang. It asks a specific question about a sphere that is still making its way into religious consciousness: what is the theological meaning of the natural world of life? This world evolved in all its splendor without human help. It was the context in which the human species itself evolved, and daily provides irreplaceable nourishment for human bodies and spirits. In our day its future is in jeopardy due to human action and inaction, destructive behavior shot through with a disastrous failure of our vaunted intelligence and virtue. As a work of theology this book explores the Christian tradition, seeking to illuminate the religious meaning of the ecological world of species. It charts one way to see that far from being simply "nature" in a neutral sense, and far from being made only for human use, these living species have an intrinsic value in their own right. Once one understands that the evolving community of life on Earth is God's beloved creation and its ruination an unspeakable sin, then deep affection shown in action on behalf of ecojustice becomes an indivisible part of one's life.

In this work I will not attempt to do justice to the vibrant contributions being made by scholars working out of religious traditions besides Christianity, nor to the advances being made in ecumenical and interreligious activity. Virtually all major religions, whether indigenous, formed in the classic axial period, or of more recent vintage, include the natural world within the scope of their vision of the Sacred. They teach a way of life that emphasizes virtues such as humility, gratitude, and compassion, and warn against vices such as pride and greed, all of which has clear implications for human behavior toward the natural world. One excellent resource for this knowledge took shape throughout the 1990s when the Forum on Religion and the Environment, led by Mary Evelyn Tucker and John Grim, sponsored ten conferences at Harvard University, one for each of ten traditions: Buddhism, Confucianism, Christianity, Daoism, Hinduism, Islam, Indigenous traditions, Jainism, Judaism, Shinto. The resulting books, named simply *Buddhism and Ecology*, *Islam and Ecology*, etc. are a goldmine of insights regarding the resources different religions bring to interpreting the natural world and promoting its flourishing.[2] Instructed and heartened by this wisdom, my own effort here remains focused on the Christian tradition with its strong belief in a creating, saving God of blessing, inherited from the Jewish tradition and now shared also with the Islamic tradition.

This book's title, taken from the biblical book of Job, reveals my starting point and operative approach. "Ask the beasts and they will teach you" (12.7), says Job; speak to the birds of the air, the plants of the earth, and the fish of the sea and they will instruct you. On the face of it, this seems a simple thing to do: consult the creatures of the earth and listen to the religious wisdom they impart. Given theology's longstanding preoccupation with the human drama, however—and we are a fascinating lot—the invitation to consult the plants and animals harbors the demand for a subtle change of method. It entails stepping outside the usual theological conversation with its presumption of human superiority in order to place a different "other" at the center of attention. The effort to approach other species with concentrated attention to their story in all its struggle and delight creates an important shift in perspective. The result changes not just what one may think about the creatures themselves, but sets up a challenging dynamic that reconfigures all of theological interpretation so that it honors their lives. All contextual, liberation, feminist, and post-colonial theologies proceed with the realization that while dominant theologies may include "the other" in some beneficial manner, the center of their intellectual and ethical interest remains the advantaged group, which does less than justice to those on the margins. The focus has to shift to those who have been silenced, so that their voices are heard and they are seen as of central importance in themselves. In a similar manner, the nascent field of ecological theology asks that we give careful consideration to the natural world *in its own right* as an irreplaceable element in the theological project.

Ask the Beasts explores this subject by conducting a dialogue between Charles Darwin's account of the origin of species and the Christian story of the ineffable God of mercy and love recounted in the Nicene creed. Given the enormous quantity of literature in both science and theology, it seemed wise to focus on one entry from each field. Darwin's book *On the Origin of Species* gives full play to life's natural character by charting its emergence through the interplay of law and chance over millions of years and thousands of miles. The Nicene creed witnesses to the gracious God who creates, redeems, and vivifies these same evolving species, grounding hope for their ultimate future. One scientific account, one religious testimony: my wager is that the dialogue between both sources, one in the realm of reason, the other of faith, can engender a theology that supports an ecological ethic of love for Earth's community of life.

The plan of this book proceeds as follows. After an initial chapter describes this project in more detail, the early chapters (2–4) focus on the evolution of species. These present background on how Darwin's extraordinary book *On the Origin of Species* came to be written; walk through its thesis in detail; and update its theory in the light of contemporary advances, lest we be dialoguing with an anachronism. A telling exchange with my colleague Terry Tilley pinpoints the importance of these chapters. He said, "You think everyone knows what evolution is, and you are bringing theological reflection to bear to connect it with faith." My heartfelt response was no, I do not think everyone knows what evolution is. I myself did not realize its ramifications before reading Darwin. For theology to have traction, we need to get the story straight. There is a parallel here, I think, with an experience common among human beings. Other persons are normally a mystery to us. Getting to know someone's story in some detail opens an avenue to greater understanding. It can move us toward appreciating, perhaps forgiving, and even loving them. In a similar manner, listening to the evolutionary story Darwin tells brings the magnificence, tragedy, and promise of the natural world into sharp relief in a way that renders it real and engaging.

With the scientific partner at the table, the middle chapters (5–8) bring the Christian story into play as this is condensed in the Nicene creed. It is a hallmark of this testimony that it professes faith in no abstract, distant deity but in its own way unspools a narrative of the living God intensely engaged with the world. Starting with the one God's creation of the world, it lingers over divine involvement with the world in the incarnation, death, and resurrection of Jesus Christ, and culminates with the vivifying Spirit who gives life and prepares a resplendent future for the whole creation. Each aspect of the story places the natural world in a different framework, all embraced by the living God. In a back and forth dialogue with Darwin's grand view of life, these chapters explore the relationship between the evolving world and one triune God.

In light of this dialogue about the beasts, birds, plants, and fish, the final chapters (9–10) turn to the human species in the grand panoply of evolution. In our day this entails facing the reality that for all our many abilities, *Homo sapiens* is ravaging the world of life. Although some still prefer to remain blind with denial, the fact is we have crossed a threshold into a new moment of human history dangerous to the well-being of the diversity of life on this planet. The novelty is captured by Aldo Leopold's awful comment, "For one species to mourn the death of another is a new thing under the sun."[3] Today thoughtful humans do mourn the disappearance of thousands of species which will never be seen again. Ruination of the earth is a deeply moral issue, as statements by official leaders of religious bodies, increasingly plentiful, make clear. Their teaching that the human vocation is to praise the Creator and care for the natural world rather than destroy it is aimed at moving their members toward what some call stewardship, or what by any name is a stance of responsibility for life on Earth.[4] Working in its own field, which may be characterized as the study of God and all things in the light of God, theology has a vital contribution to make. By uncovering the importance of plants and animals and their ecosystems in their own relationship to God, such study can invigorate ethical behavior that cares for them with a passion integral to faith's passion for the living God. In the process, human beings find their own identity reimagined as vital members of the community of creation rather than as a species divorced from the rest, and step up to protect Earth's creatures as neighbors whom they love. *Ask the Beasts* ends with this possibility as a hope, an obligation, and a prayer.

While writing this book I was somewhat daunted to discover that Karl Rahner, whose turn to the subject in theological method has greatly influenced my own thought, once declared, "The whole of Christian theology should, in the right sense of the word, be 'subjective.' It cannot speak of objects that are situated beyond the spiritual, personal, free human reality. We cannot make a theological statement about a ladybug."[5] But that is precisely what this book aims to do. It reflects on the ladybug and all its kin in the world of species beyond the human, finding them to be an intensely important if overlooked subject of religious value. To borrow a colorful metaphor proposed by Sallie McFague, in our day theology needs to sew a quilted square of its own design which, when joined to panels of other scholarly disciplines and civic activities, will be able to cover the earth with a blanket of planetary care.[6] There are many ways this can be done. The theological contribution itself will rightly be pieced together with patches constructed from diverse methods, sources, and lines of discourse. Drawing largely from the Catholic intellectual tradition, this books offers one patch for the theological square.

The last paragraph of Darwin's *On the Origin of Species* opens with a beautiful image well-known to him from many walks in the English countryside:

> It is interesting to contemplate an entangled bank, clothed with many plants of many kinds, with birds singing on the bushes, with various insects flitting about, and with worms crawling through the damp earth, and to reflect that these elaborately constructed forms, so different from each other, and dependent on each other in so complex a manner, have all been produced by laws acting around us.[7]

Note the ecological richness of his vision, with the entangled bank's plants and animals "dependent on each other in so complex a manner" and equally reliant on the soil and water of the damp bank they inhabit. I invite you, the interested reader, to keep before your mind's eye your own version of an entangled bank, whether it be an ocean beach, an urban park, a lake front or riverbank or wetland, a farm or woods, a block of city trees, a prairie or mountain range, the side of a highway or an open field, a nature reserve, a coral reef, a public garden, plantings on a campus or in a backyard garden, or even a window box on the sill—any place where land or water with their plants or animals, domestic or wild, has drawn your attention, refreshed your spirit, even lifted your mind and heart to God. Let this place function in your imagination as a touchstone for sifting through the ideas that lie ahead. We are embarked on a dialogue. The goal: to discover that love of the natural world is an intrinsic part of faith in God, to practical and critical effect.

1

BEASTS AND ENTANGLED BANK: A DIALOGUE

When we contemplate the whole globe as one great dewdrop, striped and dotted with continents and islands, flying through space with other stars all singing and shining together as one, the whole universe appears as an infinite storm of beauty.

John Muir

CREATION IN AND OUT OF FOCUS

A beautiful passage in the book of Job guides this theological exploration like the North Star that mariners steer by. The text appears when, in debate with his misguided friends, that suffering man in the land of Uz challenges them to abandon their rigid certitude about how the world works and look to another source of wisdom:

> Ask the beasts and they will teach you;
> the birds of the air, and they will tell you;
> ask the plants of the earth and they will teach you;
> and the fish of the sea will declare to you.
> Who among these does not know
> that the hand of the Lord has done this?
> In his hand is the life of every living thing,
> and the breath of every human being.

(Job 12.7-10)

1

If you interrogate the flora and fauna of land, air, and sea, the text suggests, their response will lead your mind and heart to the living God, generous source and sustaining power of their life. In their beauty, their variety, their interacting, their coming to be and passing away, they witness to the overflowing goodness of their Creator. They even teach something about human beings, that these members of the community of life also receive their every breath as a gift from the same immensely immeasureable Giver of life.

Theology, which seeks to understand faith more deeply in order to live more vibrantly, has work to do here. For in truth it has seldom asked the beasts anything. At first glance, this omission seems odd. The idea that God creates this world, not just human beings but the whole universe, is a central pillar of Christian belief. The Bible itself opens with a magnificent mythic hymn detailing how God utters the world into being with all its various habitats and fecund creatures, delighting in it on the sabbath day; and the Bible closes with a vision of a transformed heaven and earth, awash in the glory of God. In less picturesque fashion all the creeds of the church include the natural world in their confession of faith. Among them, the Nicene creed confesses the church's belief "in one God, the Father almighty, Maker of heaven and earth, of all things visible and invisible;" and in Jesus Christ, "through whom all things were made," who became part of creation through incarnation and lived all the way through to suffering and death; and "in the Holy Spirit, the Lord and Giver of life" who awakens hope for "the resurrection of the dead and the life of the world to come." There is not a catechism that does not make the doctrine of creation a central teaching. A key corollary is the intrinsically worthy quality of what has been created: "and God saw that it was good" (Gen. 1.10).

However, over the centuries for a variety of reasons which we will examine, theology narrowed its interests to focus on human beings almost exclusively. Our special identity, capacities, roles, sinfulness, and need for salvation became the all-consuming interest. The result was a powerful anthropocentric paradigm in theology that shaped every aspect of endeavor. It cast christology, for example, in its mold; the good news of the gospel flowing from the death and resurrection of Christ offered hope to human beings while the great biblical theme of cosmic redemption flew by in silence. Every area of theology can be charted making similar restrictive moves. Even the theology of creation, once it gave due play to the appropriate truths, receded to become a backdrop for the human drama. The natural world was simply there as something God created for human use. Theology lost touch with the universe.

This sketch is incomplete because there were always exceptions. In an influential study, *The Travail of Nature*, Paul Santmire uncovers how Irenaeus and the mature Augustine, among some few others, included the natural world in their understanding of the history of salvation.[1] Other historical studies find the ecological motif showing up in some of the early desert Fathers and Mothers, first-millennium Celtic saints like Bridget and Ciaran, and medieval thinkers such as Hildegard of Bingen, Meister Eckhart, and Julian of Norwich. It appeared around the edges of medieval theologians like Thomas Aquinas and Bonaventure and the reformers Martin Luther and John Calvin; surfaced in the sensitivities of John and Charles Wesley and Teilhard de Chardin; and has blossomed in Eastern Orthodox theologies. By every account Francis of Assisi stands out for his loving sense of connection and blessing of fellow creatures. This subterranean stream of creation theology is a resource that shows Christianity is not an inherently anti-ecological faith. Yet in the overall voice of theology heard in churches, universities, seminaries, and pulpits until the latter part of the twentieth century, the natural world as a subject of religious interest had largely slipped from view.

Without losing valuable insights into the grandeur, misery, and salvation of the human condition, theology in our ecological era needs to broaden its anthropocentric focus for its own adequacy. It needs to reclaim the natural world as an essential element both theoretically and in practice, or risk missing one of the most critical religious issues of our age which will affect all foreseeable ages to come. It is not a matter of either-or, of either human importance or the value of all other life. The ecological crisis makes clear that the human species and the natural world will flourish or collapse together. Given the long eclipse of interest in species other than human, however, the mandate now is to bring the buzzing, blooming world of life back into theological focus. We need to "ask the beasts."

A clarification of terms becomes helpful at the outset. In an important essay written in 1972, "A Problem for Theology: The Concept of Nature,"[2] Gordon Kaufman's analysis of the multiple meanings of the terms "nature" alerted theologians to its complex and even contradictory uses. There is nature as opposed to history; nature contrasted with grace; nature different from the polluted city; nature as a counterpart to culture; natural religion as compared to revealed religion; natural law contrasted to actions that are unnatural; and a plethora of other meanings. In ecological theology today the term tends to be given the precise meaning of the totality of processes that make up the universe and which exist independently of human beings. Even here ambiguity arises because, having evolved from this world, humans are part of it, live interdependently with it, and by our actions affect it for good or ill. What is nature? Not only is no simple definition possible, but the multiple uses of nature in different philosophies, some inimical to Christian belief, warn against a simplistic usage by theology. Two generations later Christiana Peppard's essay "Denaturing Nature"[3] updated consideration of this word's polyvalent meanings, contending that far from being a self-evident or neutral category, nature is a historically constructed idea whose meaning is affected by dynamics of power, privilege, and patriarchy. Whether as the realm of what is empirically knowable by science, as economic resource, as wilderness, as an identification of the different capacities of men and women, or myriad other meanings, the way we deploy the concept needs to be deconstructed in order to lay bare the consequences for our thinking, acting, and social arrangements.

Acknowledging the complexity of "nature," I have opted to use "the natural world" to refer to the evolving community of plants and animals and the habitats they occupy on this planet. In this book the phrase is synonymous with "the world of life" and "the living world," all bearing an ecological resonance that includes the interactions of organisms with each other and with their environment. While human beings are now interwoven with this world and are having a tremendous impact, we are late arrivals to the story Darwin tells. In the effort to give ear to the wisdom of the beasts, the usage of this book brackets the human species out from "the natural world," though not from theological consideration as part of the whole community of life on this planet.

A further distinction needs to be made between the natural world and creation. The latter is a specifically theological term whose use signals that the natural world studied by science is being viewed through the lens of religious belief. While not as multifaceted a term as nature, creation has also been employed with a diversity of meanings by different schools of thought. I use it here to refer to the living world in light of its relation to the God who creates it. Language of creation signals that this finite world is pervaded with the "absolute presence" of the living God who empowers its advance in the beginning, continuing now, and moving into the future.[4]

Bringing the natural world as creation back into theological focus at this point in time cannot be done without recognizing that we now live in a situation fraught with peril.

In our day, a new awareness of the magnificence of Earth as a small planet hospitable to life is growing among peoples everywhere. It is an ecological awareness, ecological from the Greek word *oikos*, meaning household or home. This living planet, with its thin spherical shell of land, water, and breathable air, is our home, our only home in the vast universe. It is also home to a wondrous diversity of species which interact to form networks of living ecosystems. Life abundant characterizes Earth, this jewel of a blue marble floating in a black sea of space. Perhaps life exists in some form on other planets (Mars?) or moons (Europa?) of our solar system. Since 1995 advanced means of detection have discovered hundreds of planets orbiting other stars in the Milky Way Galaxy, 892 as of June 2013, and more continue to be found. Whether any of these extrasolar planets have the potential to support simple or intelligent life or actually already do so is a matter of intense scientific interest. Definite knowledge one way or the other lies in the future. At this moment, Earth is the only place we know of in the vast universe that is "home" to living creatures.

Ecological awareness has arisen in a paradoxical context. On the one hand, we stand in awe at modern scientific discoveries about the enormous age, size, interrelatedness, and ongoing dynamism of the universe of which life on Earth is a part. On the other hand, we are struck with the terrible knowledge that we humans are inflicting deadly damage on our planet, ravaging its identity as a dwelling place for life. Both of these factors, the wonder and the wasting, shape our view of the natural world, but it is the destruction that demands attention insofar as it puts the future of life in jeopardy. The excess consumption, exploitation of resources, and polluting practices of a growing human population are dealing a sucker punch to life-supporting systems on land, sea, and air, making for nightmare headlines: global warming, melting ice caps, rising sea levels, rain forests logged and burned, ruined wetlands, garbage-filled oceans, polluted rivers, suffocating air, poisoned soils. The widespread destruction of habitats on land and sea has as its flip side the extinction of plant and animal species that dwell in them. By a conservative estimate, in the last quarter of the twentieth century 10 per cent of all living species went extinct. The dying off has become more rapid in the twenty-first century. The behavior of the human species is killing birth itself, shutting down the future of our fellow creatures who took millions of years to evolve. In the blunt language of the World Council of Churches, "The stark sign of our times is a planet in peril at our hands."[5]

The picture darkens as we attend to the deep-seated connection between ecological devastation and social injustice: ravaging of people and ravaging of the land go hand in hand. Patterns of global trade for profit that have ruinous effects on ecosystems through practices of extracting and polluting also impoverish human communities that dwell in the affected regions. In villages that survive by subsistence farming or gleanings from local forests, large-scale development projects damage both the environment and human livelihood. Lack of clean water which drives down biodiversity also takes the lives of human babies, who die in poor nations in disproportionate numbers. In urban centers economically better-off people can live in green neighborhoods while poor people are housed near factories, refineries, or waste-processing plants which heavily pollute the environment. The bitterness of this situation is exacerbated by racial prejudice as environmental racism pressures people of color to dwell in these areas. Feminist analysis clarifies further how the plight of the poor becomes exemplified in poor women whose own biological abilities to give birth are compromised by toxic environments, and whose nurturing of children is hampered at every turn by lack of clean water, food, and fuel. The ruination of habitat and the wide-scale perishing of species, with concomitant devastating effects on human beings living in poverty, intertwine in a vicious circle in rural and urban areas alike.[6]

In this context, an increasing number of superb theologians and ethicists have taken up the challenge to consider the natural world as a proper subject of attention. In their debt and in conversation with their insights, this book joins the effort to put the natural world back on the theological map. It approaches the subject by setting up a dialogue between one scientific book and one religious creed. The option for dialogue is but one among several ways theology may interact with science. Stepping back for a brief reflective scan of the field will serve to illuminate where I position the work of this book and will clarify what its method is and is not aiming to do.

MODELS OF ENGAGEMENT

Both science and theology are human endeavors which are practiced with an overwhelming wealth of methods, topics, interests and goals. An entire landscape of nuanced and unnuanced positions exists where they have intersected over the centuries. In his excellent 1990 Gifford Lectures, Ian Barbour brought some order to the scene by proposing that there are four types of science-religion interaction, namely, conflict, independence, dialogue, and integration.[7]

Science and religion may engage in *conflict*. This occurs when one or both parties transgress the boundaries of their own discipline and make claims that overlap with assertions made by the other. Each contends that only one position is legitimate, namely theirs, and rejects insights offered by the other. Evolution has been a particularly fertile field for this type of warfare. On the one hand, Christian fundamentalists interpret the biblical book of Genesis literally as if it renders an historical account of creation: God speaks and species come into existence, day after day, for six days. All creatures alive in the world were created this way. As a result, adherents of creation science or intelligent design judge evolution to be absolutely contradictory to the revealed word of God in scripture.[8] Whatever scientists have discovered can be otherwise explained. That a literal approach is not the only way to read the biblical text is roundly ignored. By contrast, contemporary biblical scholarship shows how the Genesis creation story can be read in accord with its genre, which is religious mythic narrative, and its intent, which is to teach that the one God who led the Israelites out of slavery is the universal Creator of all that exists. The sun, for example, is not divine as Israel's neighbors the Egyptians thought; rather, it is a creature, and a good one. The Bible does not aim to teach scientific facts but religious truth, "that truth which God wanted put into sacred writings for the sake of salvation," in the words of the Second Vatican Council.[9] In adhering to a literal reading of the text, the fundamentalist method of interpretation rejects this approach, thereby setting up an unyielding conflict with science.

On the other hand, operating from a different kind of fundamentalism, others seek to demolish religion with the guns of evolutionary theory. Since the theory explains the design of the living world by natural biological mechanisms, they conclude that a Creator acting as described by religious fundamentalists (and this is a key, often overlooked point) does not exist.[10] To be clear: the integrity of scientific method requires that it seek natural explanations for what occurs in the natural world. A scientist cannot properly introduce God to account for a phenomenon that is not yet understood. In that sense, scientific method is properly atheistic. The game changes, however, when thinkers allow scientific understanding of the natural world to expand into a metaphysical claim. What results is materialism, or evolutionary naturalism, the belief that "matter" is the ultimate origin and destiny of all that is. The fundamentalism here consists in taking natural explanations as the last and only word on all reality, including the phenomenon of mind. But spiritual realities, if such do exist, cannot be measured by precision instruments. Whether or not God exists cannot be resolved by scientific method, according to the definition of both God and scientific method. Materialist critics of religion thus step beyond the zone of their own expertise in making philosophical judgments about the non-existence of spiritual or ultimate reality, judgments not warranted by scientific evidence, however skilled.

Confrontation between adherents of both points of view results when the creative action of the absolute mystery of God is presented as a concrete cause among other causes in the world that science can detect ... or not, which leads to the fight. The heated public and political arguments are as fresh as today's headlines. For those not grasped by a fundamentalist commitment, however, both sides are off base in this debate.[11]

Alternately, science and religion may simply ignore each other and go about their business with a certain *independence*. Each has its own method and sources, its own field of competence, and its own proper authority. There is no overlap because each serves a very different function in human life. Science investigates the natural world to discover, control, and predict how things work. Religion articulates ultimate meanings and a path of moral behavior befitting those meanings. There need be no conflict because they are working in separate, watertight compartments. One of the most influential statements in this regard is Stephen Jay Gould's idea of the two fields as "non-overlapping magisteria," or NOMA.[12] Respectful distance is the optimum way to relate, each keeping to its distinct sphere.

There is a pressing pastoral reason why this position is not satisfactory. To be plausible to any generation, Christian faith must express itself in ways consistent with the understanding of the world available at the time. Almost by osmosis many people absorb scientific results from the intellectual culture of the day. If they are people of faith, they also adhere to teachings about the religious meaning of the world. Since very few manage to live in a mentally bifurcated world, the need for a coherent worldview becomes pressing. In a prescient observation Pope John Paul II encouraged theologians, who historically have made little effort to understand the findings of science, to take the initiative now, because:

> The vitality and significance of theology for humanity will in a profound way be reflected in its ability to incorporate these findings ... As these findings become part of the intellectual culture of the time, theologians must understand them and test their value in bringing out from Christian belief some of the possibilities which have not yet been realized ... The matter is urgent ... Christians will inevitably assimilate the prevailing ideas about the world, and today these are deeply shaped by science. The only question is whether they will do this critically or unreflectively, with depth and nuance or with a shallowness that debases the Gospel and leaves us ashamed before history.[13]

Since scientific knowledge is intrinsic to the way many people perceive the world, it is not an appendage to be set aside. Rather, as Christopher Mooney has persuasively argued, "if God is in fact the all-encompassing reality Christian faith proclaims, then what science says about nature, whether physical, chemical, or biological, can never be irrelevant to a deeper experience of God."[14] For the sake of the integrity of the truth it seeks to teach and live by, theology needs to take account of how the world created by God actually works, according to the best of our current human knowledge.

Given that both science and religion deal with the same one world, albeit differently, a third option can be for *dialogue*. Here science and religion agree that they are distinct fields of endeavor, but rather than consider the other hostile or irrelevant, they approach each other with interested respect. While they cannot answer each other's proper questions, a sharing of insights from one field to another may enrich or even correct each other. John Paul II expressed the benefit of dialogue this way: "science can purify religion from error and superstition; religion can purify science from idolatry and false absolutism. Each can draw the other into a wider world, a world in which both can flourish."[15] In fact, a dynamic interchange in which each is open to the insights of the other is necessary, he continues, so that science and religion as institutions will contribute to building up human culture, rather than fragmenting it. In dialogue each field maintains its own identity but with a permeable interface. Science informs theology's view of the very world it reflects on in the light of God. Religion offers grounding reasons why the world which science investigates is such an orderly, beautiful, coherent totality, so very comprehensible. Their exchange can provide a helpful consonance for thought and action.

Engaging in dialogue theology does not seek to prove religious tenets by appeal to scientific information. Rather than seek new evidence for religious teachings, it is on the lookout for new insight into their meaning. As theology it proceeds from its own sources of knowledge, starting with the testimony of scripture and playing through the whole tradition's witness to the self-revelation of God, while hoping to be enlightened by another source of knowledge. The conversation is based on the conviction that reason is not an enemy of faith. The ability to investigate the natural world is a gift given to the human species by the same living God who created it, the same holy Mystery who calls forth trusting faith. Copious tensions between different points of view abound as we time-conditioned humans try to figure out what is actually the case. Ultimately, however, contradictions between what reason discovers and what faith confesses are resolvable, though it may take generations. Dialogue is based on the view that the book of nature and the book of scripture, to use that lovely ancient metaphor, have the same author.

A fourth option, *integration*, is akin to dialogue but takes it a step further. Here there is a direct connection between the content of the two fields as thinkers form a deep synthesis of scientific ideas with religious belief. One example would be process philosophy and theology, an inclusive metaphysical position shaped by fundamental insights from both evolutionary science and Christian religious thought. Its insight that God is the source of novelty, immanent in the processes of the world and operating with persuasive rather than dominating power, has been widely influential, though not without critics. The thought of Teilhard de Chardin is a different illustration of this model. His scientific and religious passions fuse in "a mystic's vision of holy matter,"[16] a sense that God is working in the evolutionary world which is pressing forward towards final convergence in the Omega Point, which he identifies with Christ. In view of the ultimate purpose of the evolutionary trajectory that has produced human life, his interpretive model sanctifies human endeavor that builds the earth toward that final destiny. Teilhard's orientation of evolution to its eschatological future remains valuable, though criticism perdures that it credits the natural process with a too clear, almost linear sense of direction, and subsumes the natural world into human destiny. For all the nuance now needed, his work, poetic and pervaded with deep spirituality, has made a lasting contribution not least by integrating science with faith at a time when the two existed in watertight compartments.

There is a fifth model of interaction that I would add to Barbour's chart, namely, *practical cooperation* for the preservation of the natural world. Using scientific knowledge about growing ecological distress, many theologians have been working on the recovery of biblical and theological themes that give strong support to an ethic of environmental care. For their part, concerned scientists have urged religious communities to use their influence to motivate responsibility toward the earth. In 1990 a group of prominent scientists led by atheist Carl Sagan issued a public statement appealing to religious groups to join in preserving and cherishing the earth. Reviewing the perils to our planetary environment which might be called "crimes against creation," they wrote that:

> Problems of such magnitude and solutions demanding so broad a perspective must be recognized from the outset as having a religious as well as a scientific dimension ...
>
> As scientists, many of us have had profound experiences of awe and reverence before the universe. We understand that what is regarded as sacred is more likely to be treated with care and respect. Our planetary home should be so regarded. Efforts to safeguard and cherish the environment need to be infused with a vision of the sacred.[17]

On issues of peace, human rights, and social justice, they note, religious institutions have proved themselves strongly influential. The same bold commitment is required now to safeguard the earth. A quarter century later, noted Harvard biologist Edmund O. Wilson started his book on *The Creation* with a charming letter to an imaginary pastor of a Southern Baptist church. This was the church Wilson grew up in, though he subsequently switched his allegiance to secular humanism. While this atheist author knows his views about life's origins are diametrically opposed to those of the fundamentalist pastor, he hopes they can find common ground in the moral mandate to respect and preserve life, which is in deep trouble: "I suggest that we set aside our differences in order to save the Creation."[18] Being two powerful institutions in culture, Wilson persuades, religion and science could go a long way toward solving the problem if they joined forces, motivated by their respective belief systems.

Enemies, strangers, good friends, married partners, or co-workers: religion-and-science relationships exist in multiple forms, with numerous standpoints vociferously advocated. *Ask the Beasts* opts for the relationship of dialogue with the conviction that both science and theology are bearers of important truth about the world. They answer different questions. Science is concerned with the world as a structured system operating according to natural causes. Theology is concerned with the same world as related to God. Both open different windows onto the order and beauty of the universe, its surprising fecundity, and its suffering, death, and finitude. Building a bridge between them can have fruitful results, despite unresolved ambiguities that may remain. Since I am a working theologian and not a working scientist, this book conducts the conversation from one side of the pairing. Respecting the integrity of scientific knowledge which exists independently of religion, this book brings the resources of theology to bear in interpreting the world of life which evolutionary science describes.

THE WEIGHT OF A THEORY

For this dialogue to work, it is important to clarify what is meant by talk of evolution as a "theory." The term is frequently misunderstood in popular culture, as if evolution were an untested hunch or a guess without supporting evidence. In scientific usage, however, a theory has a stronger meaning. A theory is a coherent statement that provides an explanation for certain phenomena. It is a well-substantiated explanation of some aspect of the natural world, crafted by pulling together observed facts and known laws and interpreting them with an insightful hypothesis. Though not absolutely conclusive, it gives a reasonable account of the data at hand that can be built on by future scientists. The theory of gravity, for example, holds that physical bodies attract each other with a force proportional to their mass. It explains why apples fall down rather than up from trees, why the Earth stays in orbit around the sun, and why tides rise and fall. Similarly, that all matter consists of atoms is also a scientific theory. Its validity is supported by the periodic table of the elements, a theoretical construct that charts the structure of matter from the little atom of hydrogen to huge atoms like uranium. Their placement provides chemists with a systematic view of how materials are composed as well as a basis for predicting how they will interact.

Over time, a theory is subject to testing. If it keeps proving successful, its explanatory power gives more and more reason to think it has described an aspect of the world correctly. If incompatible evidence arises, a theory's tenets can be refined and rethought. If evidence to the contrary grows too massive, a theory can be rejected altogether. The longer the central elements of a theory hold firm—the more facts it explains, the more instances it predicts, the more tests it passes—the stronger becomes its credibility in the scientific community.

The theory of evolution is just such a successful construct. Far from being a mere speculation, it is based on a solid and growing body of empirical evidence. Its predictions have been tested with positive outcomes. Since Darwin's death, advances in science such as the development of genetics, far from invalidating the theory, have affirmed and deepened its basic tenet. At this stage there is no reasonable scientific debate about its core accuracy, only over details. Its insight into how and why a vast diversity of plants and animals have come to exist on earth, both now and in the past as revealed by the fossil record, has become a central organizing principle of the study of biology on every continent.

Thinking along these lines, Pope John Paul II garnered public interest and front page headlines in 1996 with his statement that the theory of evolution is more than a hypothesis. In 1950 his predecessor Pius XII had taught that evolution was a hypothesis worthy of study along with its alternative. Now, however, defining theory along commonly agreed-upon lines as above, the pope declared:

new knowledge has led us to realize that the theory of evolution is no longer a mere hypothesis. It is indeed remarkable that this theory has been progressively accepted by researchers, following a series of discoveries in various fields of knowledge. The convergence, neither sought nor fabricated, of the results of work that was conducted independently is in itself a significant argument in favor of this theory.[19]

Put through its paces, the explanation which evolution offers has solidified into a theory, meaning it now holds its own as a tested, convincing interpretation of the facts of life.

In light of the scientific meaning of the term, it is clear that to call evolution a "theory" is to endow it with *gravitas*. No fly-by-night opinion, it is a tested and serious explanation of how the world works, universally operative in contemporary science. Rather than spending time arguing with the critics of evolution, I accept the theory in essence as scientifically demonstrated and use its grand view of life when interpreting the natural world from the perspective of religious belief. When we "ask the beasts" their theological meaning, it is the creatures of the "entangled bank ... all produced by laws acting around us" whom we will interrogate. This is not to say "I believe in evolution." Any number of people have made this statement to me once they discovered the subject I was researching. In truth, I find the linguistic parallel with the opening line of the creed jarring. In my view, it would be more in keeping with the nature of evolution as a scientific theory to say only that one accepts it as demonstrated, and to reserve language about belief for precious human relationships and ultimately only for God.

A WAGER: GOOD DIALOGUE PARTNERS

A final introductory word about the partners which this book places in dialogue explains why I think their choice is warranted.

In his ground-breaking *On the Origin of Species* Charles Darwin brilliantly demonstrated that the variety of life on Earth has come into existence through an ages-long, complex, and astonishing history. Darwin was not the first thinker to discover that species evolve from one another, nor the only naturalist to figure out the means by which life diversifies, namely, natural selection. Yet in popular culture his name is the only one associated with evolutionary theory. Perhaps this is due to the fact that his book presented the theory of evolution to the world with such fascinating examples and rhetorical power that it could not be overlooked.

In a smart, funny newspaper essay, Carl Safina deplored this connection, arguing that for evolution to be accepted we need to "kill Darwin." Consider how far the theory of evolution has come since Darwin's time. Though a brilliant thinker, he wrote before science had knowledge of genetics or DNA, and before the development of molecular biology. In the more than 150 years since the publication of *Origin*, thousands of scientists have challenged and tested his ideas in field work and labs, adding vast new knowledge to how evolution works. We do not call gravity "Newtonism," but a law of nature attested to on many fronts. Calling evolutionary theory simply "Darwinism" makes it into something of a cult idea to be believed in or contested:

> By propounding "Darwinism," even scientists and science writers perpetuate an impression that evolution is about one man, one book, one "theory." The ninth-century Buddhist master Lin Chi said, "If you meet the Buddha on the road, kill him." The point is that making a master teacher into a sacred fetish misses the essence of his teaching. So let us now kill Darwin.[20]

The idea of evolution would have come on the scene, Safina argues, with or without the work of this naturalist. It stands on its own strong legs by now. Let it swim in the public pool of ideas without the magic talisman of Darwin's name.

And yet! *On the Origin of Species* is arguably one of the most significant books of the modern era, a genuine game-changer in the history of ideas. The book itself is a scientific classic. It marks a new epoch in the development of the natural sciences, establishing the view that life-forms previously thought to be permanent are actually always in flux, coming into being and passing away in close relation to one another. The pattern of thinking in many other disciplines and in culture itself changed after this. More than that, *Origin* is beautifully written. To speak personally, I have found it a fascinating read, several times over. The loving sensibility of Darwin's observations, the sharpness of his logic, the strength of his synthesizing, the power of his persuasive rhetoric, the charm of his honest confessions of ignorance, the telling inclusion of arguments raised against his own position (so reminiscent of Aquinas), the wit against opposing views—all combine with sweeping power to deliver new insight into the dynamic world of life. If theology is to engage with the natural world at all, it could hardly do better than to entertain the vision of this book.

As for the Nicene creed, composed by two church councils in the fourth century, Nicea in 325 and Constantinople in 381, and now a bond of unity (more or less) among Christian churches around the world, it too may seem dated. At the very least its language speaks with the vocabulary of a bygone era, and in recited practice it seems to be got through without much attention or enthusiasm. Yet pulsing underneath its three-fold structure is a narrative of divine engagement with the world, a story crystallized in brief phrases filled with promise. There is one God, congregations declare and choirs sing, who creates, becomes one with, redeems, and makes holy the world, now and into the hoped-for future. Such in shorthand is the living God I allude to in this book. While natural philosophy will have a role in elucidating certain concepts, I am not talking about divinity inductively arrived at in the abstract. Rather, I am reflecting squarely on the Creator God, God of the Exodus and covenant with Israel, the God made known in the life, death, and resurrection of Jesus Christ, the God whose Spirit vivifies all things to heal, redeem, liberate, and ultimately to renew the whole creation. Ineffable holy mystery, this is the one trinitarian God of Christian faith, made known through scripture and the living tradition of creed, doctrine, prayer, active witness, and the religious experience of believers up to this day. Given this explicitly Christian location, my search for understanding finds its center in the gospel, whose teaching can be capsulized in the phrase "God is love" (1 Jn 4.16). God is the incomprehensible mystery of love beyond imagining. This insight, itself a summary of what the creed confesses, will guide the explorations that lie ahead.

Asking the beasts, birds, plants, and fish about their evolutionary relationship with the God who is love brings forth a multifaceted answer. Matching their response with the creed read in reverse we hear:

❧ "We are fecund and exuberantly alive." The spark of life is kept flaring by the Holy Spirit, Giver of life, the vivifier who dwells within all things empowering their advance. Unlike Orthodox theology of the Eastern churches, Western theology until recently had little to say about the intimate nearness, the immanence, of the Spirit of God's presence in the cosmos. Neglect of the Spirit and of the religious value of the natural world seem to go hand in hand. Classical theology, however, when it did consider the Holy Spirit, figured the Spirit's proper name to be love. The reflections ahead aim to explore a vigorous pneumatology, placing the fire of divine love at the center of the evolutionary world. Far from over-riding the beasts' natural processes, this love empowers and sustains them in their evolving autonomy.

❦ "We suffer and die." Given the enormity of death and extinction entailed in evolutionary history, this dialogue will inquire about divine solidarity with creaturely suffering, enacted most clearly in the cross of Jesus Christ. Theology has yet to plumb the depths of the meaning of incarnation, "the Word became flesh" (Jn 1.14), for the natural world, or the fullness of the meaning of the resurrection of the crucified for the whole of creation. Framed by these doctrinal symbols, the mercy embodied in the ministry of Jesus unto death provides a key to divine compassion toward the beasts.

❦ "We are created." Ultimately the beasts do not ground themselves, but receive their existence as a continuous gift from the living God who is the Creator of all. Their eschatological transformation in glory is also beyond their potential, but is a promise embedded with the gift of their life. In community with the rest of creation, species become known as treasured beneficiaries of the gracious, life-giving God who is *Alpha* and *Omega*.

"There is grandeur in this view of life," Darwin wrote in the beautiful last sentence of his spectacular book (490).*

There is also grandeur in the view of life presented in the creed, this same evolving life embraced by the living God of love. This book's dialogue between these two views hopes to offer a rich fare of challenge and insight to contemporary theology and the life of faith and morals it seeks to serve. Knowing the evolutionary story in some detail can open vistas of appreciation and deep wells of compassionate fellow-feeling toward other creatures. Seeing them embraced by the triune God can crystallize their identity as a creation of inestimable value, replete with religious significance.

* All citations are from the first edition.

From this base can flow compelling reasons to fiercely prize biodiversity and act responsibly for the care of all living species. Not incidentally, such engagement can also deepen theological understanding of the ineffable God's creative and redeeming action in the world, which would be Darwin's "gift to theology," as John Haught has audaciously claimed.[21] Grandeur enough for all.

2

"When We Look ..."

To stand at the edge of the sea, to sense the ebb and flow of the
tides, to feel the breath of a mist moving over a great salt marsh,
to watch the flight of shore birds that have swept up and down
the surf lines of the continents for untold thousands of years,
to see the running of the old eels and the young shad to the sea,
is to have knowledge of things that are as nearly eternal as any
earthly life can be.

Rachel Carson

THE AUTHOR AND HIS AMAZING BOOK

It took less than one year for Charles Robert Darwin to write
On the Origin of Species, published in London on November 24,
1859. Not that he started from scratch. For several years prior he
had been hard at work on his "big species book," a massive treatise
that would give evidence for his theory that species descend from
other species via a branching process governed by natural selection.
This book was still years away from completion when in June of
1858 the threat of competition shifted him onto a faster track.
He received a manuscript in the mail from the British biologist
and geographer Alfred Russel Wallace, then working on the island
of Ternate in what is today Indonesia. Reading it, Darwin was
astounded to see that Wallace had arrived at similar ideas about

the origin of species, including the role of natural selection. Thinking that his own originality would be questioned, he quickly consulted with friends who urged him to go public. The next month, extracts from Darwin's writings were read out at a meeting of the Linnaean Society, along with Wallace's paper; both were published to immediate interest. To forestall further scoops, Darwin decided to write an "abstract" of his unfinished larger work on the subject. This "abstract," fully titled *On the Origin of Species by Means of Natural Selection, or the Preservation of Favoured Races in the Struggle for Life*, stands as a groundbreaking work in the history of Western culture, remarkable for both its scientific and literary merit.[1]

While it took him nine months to write this book, all of Darwin's 50 years of life up to that point actually prepared for it. Born into a wealthy family of privilege with strong intellectual interests, he was well schooled in natural history as a youth. His renowned paternal grandfather, Erasmus Darwin, had already written favorably about evolution, advocating in addition the abolition of slavery, education for women, and other forward-looking views. Josiah Wedgwood, his maternal grandfather, a prominent abolitionist and business entrepreneur, was keenly interested in the scientific advances of his day and used its methods to revolutionize the manufacture of quality pottery and china. There was family precedent, encouraged by his freethinking father, for welcoming rather than shutting out the challenge of new ideas.

Darwin's young adulthood continued his remote preparation for the book. Five years of college, followed by five years traveling, and then five years interacting with the scientific community in London led him to carefully honed insight about the dynamic forces that shape the natural world. Throughout this time his thinking was deeply influenced by the British tradition of natural history and natural theology, and inseparable from the soil of British social, commercial, and political life.[2]

His college education started with two years at Edinburgh University where, following in his father's and grandfather's footsteps, he studied to be a physician. Finding the medical lectures dull and the clinical practice horrifying, especially peoples' pain under bloody surgery with no anaesthetic, he preferred to attend lectures in chemistry, geology, and natural history which exposed him to contemporary debates about the history of the earth and its fossils. The lad had a precocious scientific interest; his free time was spent exploring marine life along the nearby Scottish coastline. It became clear that medicine was not for him. His disappointed father sent him next to Cambridge University for a degree that would prepare him to become a priest in the Anglican church. There he lucked into mentors who taught him an approach to nature through precise methodology: meticulously observe the particular animal, plant, or rock in its broad context, and let this observation lead by active induction to theorizing on a grand scale. Once again, less than enthralled with the study of divinity, he joyously pursued his passion for nature in extra-curricular local excursions with professors to study botany and geology.

In summer of 1831, shortly after he graduated from Cambridge, Darwin received the now famous invitation to travel aboard the *HMS Beagle* as a self-paying, unofficial naturalist and dining companion to its captain, Robert FitzRoy. Sponsored by the Naval Admiralty Office, the expedition had the goal of charting the southern coast of South America for purposes of promoting the British empire's commercial and military interests.[3] Originally planned for two years, the survey lasted five (1831–6) and circumnavigated the globe. Darwin spent more than three of these years on land, making inland expeditions, studying geologic formations, digging out fossils, observing the native flora and fauna, and collecting specimens of living species. Back on ship when he wasn't fighting seasickness, he had time to read, reflect, and compose extensive diaries with careful, copious notes and drawings of his observations of the land, its fossils and living creatures, and how they all fit together. What a boon to a young, curious mind! Although not a trained geologist or biologist, his education had prepared him well with basic knowledge and methods of studying the natural world. Still, he found himself applying, testing, and modifying inherited theories against a set of personal experiences that far transcended those of his teachers. The first lines of *Origin* written some twenty-five years later testify to the importance of this trip for his thinking:

> When on board the H.M.S. 'Beagle,' as naturalist, I was much struck with certain facts in the distribution of the inhabitants of South America, and in the geological relations of the present to the past inhabitants of that continent. The facts seemed to throw some light on the origin of species – that mystery of mysteries ... (1).

Thanks to specimens, letters, and journal segments periodically sent home during the journey, Darwin arrived back in England to a warm welcome in scientific circles. Neither a doctor nor a clergyman would he be. Instead, he lived his next five years in London as a city gentleman of science (his father set up a fund), reaping the intellectual benefits of his trip. Joining professional societies, reading voraciously, attending scientific meetings, and engaging in intense conversation with members of the scientific guild, he was nourished by and contributed to the ferment of ideas about the origins and relationships of species that so engaged the interest of the intellectual community. Early on he published a highly popular account of his adventures now named in shortened form *The Voyage of the Beagle*, both an exciting travelogue and a detailed field journal, laced with his growing theories about the relationship of species over time.[4]

More significantly, he kept a series of small, leather-bound notebooks on earth, life, and mind which charted the growth of his thinking about a law that might govern the succession of species in time. The anatomist Richard Owen judged that the fossil bones Darwin had collected were those of extinct creatures, but these were apparently related to species now alive above ground in South America. One prize fossil specimen, for example, was a large, armored mammal given the name *Glyptodon*; its skeleton bore unmistakable similarities to the modern armadillo. Both animals were found in Latin American and nowhere else. Darwin was soon speculating in his Red Notebook on the possibility that within the same area "one species does change into another" to explain this affinity. The ornithologist John Gould announced that the Galápagos birds which Darwin had thought a mixture of blackbirds, gros-beaks, and finches were, in fact, 12 separate species of finches. Moreover, these species were similar not to birds of other rocky oceanic islands around the world, but to those of the closest large landmass, the South American continent. This was so, puzzlingly, even though the lush tropical conditions of the mainland could hardly have been more different from the arid conditions on the off-shore islands. These islands were of young and volcanic origin. Might it be that ancestor finches migrated from the mainland, and then diversified due to different environmental conditions? Could it be their descent from a common ancestor that explains so much similarity in difference?

Darwin mapped this momentous idea not just onto birds but onto all of life, from the earliest creatures up to human beings, sketching a little branching tree to catch the idea. By winter 1838–9 the theory of natural selection was fully formulated in his London notebooks. These five years were the most intellectually fertile period of Darwin's life. In truth, the London notebooks contain almost all the theoretical insights he would later publish.[5]

It was during these city years that Charles wooed and wed his first cousin Emma Wedgwood, from all descriptions a charming, intelligent, religious woman, forming with her a marriage of life-long mutual support. With two little children in hand and a third on the way, they bought a house and some 20 acres of land in Kent about 16 miles southeast of London. Here they remained for the rest of their days, living the life of the landed gentry. Biographers note that it was Emma's energetic abilities that ran the household, held their family together, and provided the stability and support her husband required to pursue his research and writing. She also stood guard during his recurring bouts of ill health. Despite his growing religious doubts they remained open with each other; letters and journals give glimpses of the consolation their affection afforded. For his part, Charles spent his days intensely occupied with close studies of the natural world: barnacles, pigeons, orchids, earthworms. He was far from being isolated, however. Thanks to the growing reach of the British postal service, he corresponded voluminously both nationally and internationally with other investigators, exchanging information, books, specimens, and, most of all, ideas about the workings of nature.[6] Thinking of himself as a philosophical naturalist, that is, a scientific student of natural history, he not only observed but theorized, publishing books and monographs about his discoveries.

The Darwins had ten children. As a Victorian *paterfamilias* Charles was by all accounts a devoted father, uncommonly attentive, worrying that any childhood illnesses might be due to biological inbreeding (his mother and Emma's father were siblings). The children were in and out of his study; he wrote about how they monitored his experiments in the garden, greenhouse, and surrounding fields, and helped in his research: "My son made a careful examination and sketch for me of a dun Belgian cart-horse with a double stripe on each shoulder and with leg-stripes" (164). These parents suffered a terrible blow when their second child, Annie, a delightful daughter, died at the age of ten in 1851, likely of tuberculosis. Desolate, Charles penned a ten-page memorial, lamenting, "We have lost the joy of the Household, and the solace of our old age. ... Oh that she could now know how deeply, how tenderly we do still & and shall ever love her dear joyous face." As Adam Gopnik notes, nothing really readies us for the sudden loss of someone we love:

> Darwin had lost his father and his mother, but nothing could have prepared him for losing Annie. It is like watching someone sink straight down into the waves, who will never return and never be recovered, while life continues on the surface. This sense ... of a life going on of which Annie no longer knows a thing, and in which her absence is absolute and permanent, is true grief – no memory can help it; no promise of meeting after can alter it. King Lear's "never's" are the horrible truth; once she was here, and she will never be again.[7]

The bereft father grieved her all his days. Two other children died as babies, one during that game-changing summer of 1858 when Wallace's paper arrived. Distraught at the death of his toddler son, also named Charles, Darwin did not attend the critical meeting where his and Wallace's papers on natural selection were first read in public.

It was in this setting, in the comfortable study of Down House, that Darwin pulled together his "abstract," drawing from notebooks and his unfinished treatise to create *On the Origin of Species*.[8] Already in 1844 he had drafted a sketch of his theory and given it to his wife to publish after his death should he not live to write an entire book. Now the idea took shape in a carefully crafted, beautifully written argument. Although quick in coming, *Origin* did not spring like Athena out of the head of Zeus. Rather, it poured forth as the ripe fruit of decades spent observing and thinking about the history of the land and the sea and the living organisms that inhabit them. As the book's "Introduction" observes, "I have not been hasty in coming to a decision" (1).

November 24, 1859. Spurred on by pre-publication buzz and group sales to lending libraries, the initial printing of 1,250 copies quickly sold out, if not exactly overnight then shortly thereafter. A second edition soon appeared, and the book has never been out of print since. While *Origin* is a rigorously argued treatise, written for scientists and quoting copiously from their experiments, books, and letters, it is also a wonderfully accessible text. General readers warmed to its personal style, courteous and persuasive tone, jargon-free language, reams of interesting examples, and lucid writing, at times lyrical and laced with creative metaphors. Many were deeply interested in the religious and cultural implications of Darwin's ideas, especially as they applied to human beings and society, in addition to his insights into nature. The book flew off the shelves.

By the next spring *Origin* had been printed in New York and thence made its way around the English-speaking world as an immediate bestseller. Darwin is certainly the most translated scientific author of all time. Starting with the German and Dutch editions in 1860, *Origin* was rendered into the French, Russian, Italian, Swedish, Danish, Polish, Hungarian, Spanish, and Serbian languages during his lifetime. It now appears in over forty languages, most recently Tibetan. A recent count estimates that approximately 75,000 to 100,000 copies of the *Origin* in book form are sold annually in various languages throughout the world, along with numerous shortened versions and commentaries. The full texts of its various editions are also available for reading online, for download as audio books and e-books, and in Braille. It may well be, as one scholar has suggested, that Darwin is the most discussed writer in English after Shakespeare.[9]

Quite an "abstract."

In the years that followed its original publication, the *Origin* itself evolved in relation to major lines of criticism. Adding explanations, omitting material, and clarifying points, Darwin kept on revising the text. In the fifth edition the fateful phrase "survival of the fittest" first appeared. Persuaded by Alfred Russel Wallace among others, Darwin took the phrase from the philosopher Henry Spencer's idea about social progress among human beings and incorporated it into his own discussion of the biological struggle for existence. Strange as it may seem, Darwin introduced the term 'evolution' only in the sixth edition. All told the book went through six different editions, the last being published in 1872. He never did finish his "big book."[10]

Not one to rest on his laurels, in the post-*Origin* years this student of nature produced a prodigious number of publications, all extending the evidence for the theory one way or another. His last book, published the year before his death, was on earthworms and their role in the formation of vegetable mold. Such a topic may seem anti-climactic, even whimsical, but as David Reznick astutely observes, "it represents a different form of his interest in how seemingly small, everyday processes can cause great change if they persist over long intervals of time."[11]

It is of historic as well as human interest to note that during these years Alfred Russel Wallace, the co-discoverer of evolution by natural selection, publicly defended his and Darwin's ideas with strong argumentative ability. At the time his original paper reached Darwin, he was a younger, less well-known, and less supported scientist working in the Malay Archipelago far away from home. When he found out that Darwin's friends had read his paper before the Linnaean Society, he professed himself to be delighted to be held in such high esteem by the eminent scientific men of London. In the *Origin* Darwin gives him honorable credit, explaining on the first pages how Mr. Wallace "has arrived at almost exactly the same general conclusions that I have on the origin of species" (2) and telling the story of how he received "Mr. Wallace's excellent memoir" (2). Throughout the book Darwin makes appreciative references to Wallace's researches, at one point citing Mr. Wallace's fundamental principle that "every species has come into existence coincident both in space and time with a pre-existing closely allied species" (355). While Darwin's name is associated with the theory of evolution, biographers indicate that his younger colleague did not feel wronged in any way. Indeed, when in 1889 Wallace came to write his own book on evolution, he entitled it *Darwinism*.[12]

After a lifetime of carefully observing the entangled banks of life large and small, Charles Darwin died on April 19, 1882. Calling him "the most widely known of living thinkers," the obituary in the *New York Times* noted that though many had not read him, "the shock of the new idea" he proposed reached the "thought of the masses until the slightest allusion to Darwinism was sure of instant recognition from even the most illiterate individual or audience." His impact on science and culture meant that "Mr. Darwin, therefore, may be called an epoch-making man." Amid an outpouring of international recognition and respect, he was buried at Westminster Abbey next to another great English scientist, Isaac Newton.

CORE INSIGHT

Darwin did not speculate about how life originally began on Earth. His thinking begins when life is already up and running, expressed in extraordinary varieties of plants and animals interacting with each other. How do all the different species that make up this beautiful world arise? Darwin proposed that all living beings originate through entirely natural processes. In face of the widespread scientific and religious assumption that species come into being independent of each other by separate acts of a divine Creator, and the view that they remain immutably themselves throughout their existence, *On the Origin of Species* is one long argument that species are in motion, coming into being from previous species by a process that can be explained naturally, without appeal to a supernatural cause. Of the book's fourteen chapters, the first five present the core argument; the next three answer difficulties; the next five show its explanatory power; and the last summarizes.

In a nutshell, Darwin figured out that species result from a chain of events that goes something like this. Living organisms produce variations (we now call them mutations). These variations can be inherited. Some variations increase an organism's ability to survive and reproduce. Given large numbers of offspring generated by a species, there is constant competition for the resources needed to sustain life. Any variation that gives an organism an advantage in this struggle will enable it to eat and mate with more success and will likely be passed onto its offspring. For example, a variation may increase the toughness of a bird's bill ever so slightly. If the area's major food source is a hard seed, the tougher bill will result in more nourishment for the bird and more successful egg-laying. Such adaptation does not happen to individuals in isolation, but as they co-adapt in relation to others and to the physical environment. Those variations which give a relative advantage to organisms in their effort to survive and reproduce go onwards into the next generation, eventually spreading through a whole population. Those that do not, die out. Over eons of time, new species diverge from ancestral parents as a result of this process.

In the first five editions of *Origin* Darwin did not call this phenomenon "evolution," but rather "descent with modification." Species share a common parentage that can be traced back in time (descent); over time they undergo changes which improve their success in staying alive and generating offspring, or not (with modification). While the initial occurrence of variation is unpredictable, the process as a whole is not haphazard but is governed by certain principles that have the character of laws of nature. One law above all is largely responsible for the outcome. This is what Darwin and Wallace had both discovered. To this main though not exclusive driver of evolution Darwin gave a name: "This preservation of favourable variations and the rejection of injurious variation, I call Natural Selection" (81). In poetic fashion he tended to personalize this principle:

> It may be said that natural selection is daily and hourly scrutinising, throughout the world, every variation, even the slightest; rejecting that which is bad, preserving and adding up all that is good; silently and insensibly working, whenever and wherever opportunity offers, at the improvement of each organic being in relation to its organic and inorganic conditions of life. We see nothing of these slow changes in progress, until the hand of time has marked the long lapse of ages, and then so imperfect is our view into long past geological ages, that we only see that the forms of life are now different from what they formerly were. (84)

Like all metaphors, this likening of natural selection to human scanning and choosing illumines the subject. But like all metaphors it also contains a strong note of dissimilarity. Unlike human actions, natural selection does not operate with conscious, intelligent pre-planning and intent. It is not an active agent. In a true sense, it is not literally a "selection" at all, as if it were a deliberate force working to eliminate the maladapted and allow multiplication of the successful. Akin to what Isaac Newton figured out to be the law of gravity, it can be called a law of nature. Planets move in orbit around the sun; an asteroid flies within the gravitational pull of a planet and goes flaming down into its surface; a full moon pulls up a great high tide in the ocean; an apple falls from a tree and a scientist sitting underneath gets bonked on the head. There is no thought or willing in any of these happenings. It is simply the way the world works according to the law of gravity. Darwin's key insight is that species originate by the working of just this kind of a natural principle. Life in all its variety, anomalies, and exquisitely adaptive design can be largely explained by natural selection operating within historical circumstances. This history is larded with struggle, impermanence, imperfection, and chance. Darwin knows he operates under a cloud of "profound ignorance" regarding so many of its aspects. Yet the closing lines of the "Introduction" to *On the Origin of Species* assert conviction about one thing:

> Although much remains obscure, and will long remain obscure, I can entertain no doubt, after the most deliberate study and dispassionate judgment of which I am capable, that the view which most naturalists entertain, and which I formerly entertained – namely, that each species has been independently created – is erroneous. I am fully convinced that species are not immutable ... Furthermore, I am convinced that Natural Selection has been the main but not exclusive means of modification. (6)

The term "natural selection" may indeed be a metaphor, a creation of language. But, as E. O. Wilson insists, "what it represents is real, and very powerful."[13]

SCIENCE: SPECIAL ACTS OF CREATION

Note the reference in the above quotation to "most naturalists." A stereotypical view holds that Darwin's theory was greeted with complete rejection by the religious establishment and unanimous support by the scientific community, in black and white fashion. This view gains some traction from one memorable, almost mythic, exchange during the well-attended first prominent debate on Darwin's theory held at a meeting of the British Association for the Advancement of Science in Oxford 1860. Ridiculing evolutionary theory, the eloquent Lord Bishop of Oxford, Samuel Wilberforce, baited biologist Thomas Henry Huxley, who had "nailed his colours" to Darwin's mast, with the question of whether he thought himself descended from an ape on his grandfather's or his grandmother's side. In true bulldog fashion Huxley gave a passionate defense of Darwin's theory, retorting in the end that he would rather have an ape for an ancestor than a man of versatile intellect who introduced ridicule into a grave scientific discussion (presumably the bishop; accounts vary and exact words are unavailable).

It is a mistake, however, to assume that all scientists were united behind Darwin and that all Christian thinkers lined up against him.

The briefest research shows that the situation was much more interesting. While some church people indeed found Darwin's idea disquieting and a threat to their religious belief, a number of theologians in Germany, France, and England endeavored to incorporate evolutionary perspectives into their work.[14] Even at that Oxford debate there were other clergymen more open to Darwin's naturalism, including Frederick Temple, later the Archbishop of Canterbury. A decade later John Henry Newman, a leading Anglican churchman who had converted to Roman Catholicism, on being asked whether Darwin should be awarded an honorary degree from the University of Oxford, reflected on evolutionary theory: "Is this against the distinct teaching of the inspired text? If it is, then he advocates an Antichristian theory. For myself, speaking under correction, I don't see that it does – contradict it."[15]

What scrambles the picture even more is the fact, made clear throughout *Origin*, that not only religious believers but most naturalists held that each species had been independently created and remained fixed and unchanged throughout its existence. Unlikely as it seems to twenty-first-century readers, most leading scientists at that time were proponents of the idea that species originated by a direct act of divine agency. As biologist David Reznick writes, "Today we think of the advocates of special creation as representing nonscientific, religious opponents to evolution. In Darwin's day, they were the scientific establishment. Virtually everyone, ranging from his professors at Cambridge to all those who had the greatest influence on Darwin's intellectual development, advocated some form of special creation."[16] Philosopher Michael Ruse even thinks that Darwin's reason for the long delay in publishing, from the private sketch in 1844 to the public book in 1859, is that "he was scared." Of whom? "It was precisely the leaders of his scientific set – those very men who had nurtured him and made his early career possible – whom Darwin feared offending."[17]

Before Darwin's book Richard Owen, the famed anatomist, had already advanced a somewhat veiled evolutionary theory, one, however, that incorporated divine design. He immediately wrote an anonymous, nasty review of the *Origin* calling it an "abuse of science" because, among other things, it prematurely jumped to a mechanism of change (i.e. natural selection), ignoring Owen's own position that the anatomy of plants and animals was a physical expression of a divine plan. In his words,

Darwin discounted the "axiom of the continuous operation of the ordained becoming of living things," ordained by an intelligence that ordered the universe. Just as bad, in Owen's view, the book's pleasing style, artistic flair, and sequence of arguments made such a good case for the transmutation of species that it "seduced" perhaps the majority of younger naturalists to this position.[18] Darwin's friend Charles Lyell, a brilliant geologist whose influential *Principles of Geology* he had first read on the *Beagle* and used in mounting his argument, believed that while each species was exquisitely adapted to its place of origin and could extend its range to some degree, its ability to produce variations was limited and did not lead to new species: "For Lyell, new species arise independently of any others, as special independent creations."[19] Three of the best-known theorists of scientific method in the mid-Victorian era, John Herschel (Darwin's teacher at Cambridge), William Whewell, and John Stuart Mill, maintained that genuine science had to be based on extensive evidence from which theory could be induced. It was legitimate to explain the origin of species by general laws without recourse to interventions of divine power in each particular case. Nevertheless, there came a point where this would be insufficient. As Whewell wrote regarding causal chains that stretched back in time, "we must contemplate supernatural influences as part of the past series of events, or declare ourselves altogether unable to form this series into a connected chain."[20] Such references to supernatural causes in the history and philosophy of science were not in the least unusual at the time. As for Darwin's theory of natural selection, each of these three concluded in his own way that, as Whewell said, "At best it was not good enough, and certainly not as credible as the theory of creation by a designing intelligence. At worst it was not a legitimate scientific theory at all."[21] Darwin found his teacher's disparagement "a great blow."

What shaped the intellectual context of the reactions of these and other noted scientists was the dominance of the argument from design in traditional British natural philosophy and theology. The logic of this argument holds that from the order and beauty of the natural world one can infer the existence and attributes of a divine Creator. The exquisite fit and functions of organisms give compelling evidence that they were fashioned by a supernatural intelligence and power. In other words, design implies a Designer. The most celebrated advocate of this position, William

Paley, promoted the watchmaker analogy in his influential book *Natural Theology, or Evidences of the Existence and Attributes of the Deity Collected from the Appearances of Nature*, published in 1802. If while crossing a heath he happened to find a watch upon the ground, even though far from a settled town, Paley would reasonably surmise that an intelligent artisan existed who had designed and constructed it for the purpose of telling time. Just as the existence of a human watchmaker can be inferred from the discovery of a watch, so too can the existence of a divine Creator be inferred from the beautiful functioning of the natural world. In fact, even more so, since every indication of contrivance, every manifestation of design, which existed in the watch exists in the works of nature to a degree which exceeds all computation. Living organisms with their many parts working together could no more come into being without a purposeful intelligence than could ticking watches. Hence, the complex structures and exquisite adaptive traits of plants and animals, such as a bird's feather with its marvelous "apparatus of critchets and fibres, of hooks and teeth," even down to the wings and antennae of the humble earwig, show the existence of an intelligent Designer.[22]

A new wrinkle was added to this argument by the discoveries of geologists who found that new kinds of species appeared successively in the fossil record of Earth's geological layers. Explorers, too, came upon volcanic islands that were relatively recent in origin but filled with life. This must mean that the divine creative power was operational not just once, "in the beginning" as told in the Genesis narrative, but continuously over time as the need arose. Special acts of divine creation were ongoing across the planet, even to the tune of "ten thousand creative acts" which the geologist Adam Sedgwick purported to find.[23] Even if one admitted a limited role for evolutionary development, the history of life came about by direct divine agency and unfolded according to a plan conceived in the mind of God.

This argument from design was deeply familiar to Darwin's audience, both scientific and lay. The popular version of the watchmaker analogy was widespread; Darwin himself had studied and been examined on Paley's work at Cambridge. It gave a compelling explanation of the order and beauty of the natural world that was scientifically respectable while it cohered with religious belief. With his notion of the origin of species by natural selection, however, Darwin had in hand a purely natural account for the structure and habits of animals and plants and, by implication, human beings. Species were mutable, not permanent; to explain their beautiful design one need only allude to natural causes, not invoke separate, sequential acts of creation by a divine Creator. Of the intellectual power of his discovery he later wrote: "the old argument from design in nature, as given by Paley, which formerly seemed to me so conclusive, fails, now that the law of natural selection has been discovered."[24]

Time and again in the *Origin* Darwin contests the prevailing scientific view that species originate by separate divine acts of creation, reasoning against it with the vigor one uses in trying to reshape the governing paradigm of a whole field of study. Using various expressions, he lasers in against those who argue for "a special act of creation" (55); "continued creation of new organic beings" (95); indigenous plants being "specially created" (115); each species being "independently created" (129); blind cave animals being "separately created" (138); the Swedish turnip, ruta baga, and common turnip having enlarged stems due to "three separate yet closely related acts of creation" (159); the upland goose still having webbed feet because "it has pleased the Creator" (185–6). His words sound a steady drumbeat of criticism that resounds at least four dozen times, by my count. He was doing what scientists always do, which is to discriminate between alternative explanations. Because the prevailing scientific belief was that each species came about by an independent act of creation, this is the position he was trying to disprove as he argued his theory. In virtually every instance he argues that the direct-creation theory offers no intellectually satisfying *explanation* of the appearance and location of species, compared with descent with modification:

> How inexplicable on the theory of creation is the occasional appearance of stripes on the shoulder and legs of the several species of the horse-genus and in their hybrids! How simply is this fact explained if we believe that these species have descended from a striped progenitor, in the same manner as the several domestic breeds of pigeon have descended from the blue and barred rock-pigeon! (473)

Darwin notes that whole orders of species are missing from oceanic islands, frogs, for example. Their absence can be explained by natural selection, for these organisms cannot survive seawater travel, and thus cannot move from continents to lands off-shore. "But why, on the theory of creation, they should not have been created there, it would be very difficult to explain" (393). Dozens of times the explanatory power of his alternative theory is emphasized:

> On the view that each species has been independently created, with all its parts as we now see them, I can see no explanation. But on the view that groups of species have descended from other species, and have been modified by natural selection, I think we can obtain some light. (152)

In standard works of natural history imperfect and useless rudimentary organs are said to be specially created for the sake of symmetry; "but this seems to me no explanation, merely a restatement of the fact" (453). Authors of the highest eminence seem to be satisfied with the view that each species has been independently created. But:

> To my mind it accords better with what we know of the laws impressed on matter by the Creator, that the production and extinction of the past and present inhabitants of the world should have been due to secondary causes, like those determining the birth and death of the individual. (488)

As for that upland goose with webbed feet, to say it pleased the Creator "seems to me only restating the fact in dignified language" (186). Those who believe that each equine species was independently created assert that the tendency to vary in particular ways is also the result of their being specially created; but:

> To admit this view is, as it seems to me, to reject a real for an unreal, or at least for an unknown cause. It makes the works of God a mere mockery and deception; I would almost as soon believe with the old cosmogonists that fossil shells had never lived but had been created in stone so as to mock the shells now living on the sea shore. (167)

Contrary to the theory of special acts of creation which simply begs the question, the theory of common descent with modification is a true reason, a *vera causa*. This term, originated by Isaac Newton in 1687, was generally understood to mean "causes recognized as having a real existence in nature, and not being mere hypotheses or figments of the mind."[25] The power of natural selection is actually at work in the real world, Darwin argued. It can be found and tested by empirical observation. So great is the explanatory power of this idea that he can simply declare: "Descent is the hidden bond which naturalists have been unconsciously seeking, and not some unknown plan of creation" (420).

The rhetorical structure of these and numerous other instances makes it clear that Darwin is arguing not first and foremost against religious belief, though he is not unaware of implications in that direction, but against a prevailing scientific paradigm.[26] While he recognizes that "[a]uthors of the highest eminence seem to be fully satisfied with the view that each species has been independently created" (306), his vigorous argument seeks to demonstrate that such a position is inadequate. It actually *explains* nothing. He acknowledges how rash it feels to differ from the great authorities: "all the most eminent paleontologists, namely Cuvier, Owen, Agassiz, Barrande, Falconer, E. Forbes, &c., and all our greatest geologists, as Lyell, Murchison, Sedgwick, &c., have unanimously, often vehemently, maintained the immutability of species" (310); he names Forbes, Pictet, and Woodward as "strongly opposed to such views as I maintain" (316). Yet despite a formidable *Who's Who* of opponents, common descent with modification by natural selection can account for a wide array of phenomena in an intensely more satisfactory manner. This argument was pitched primarily at his professional peers. For all his vocal supporters such as Huxley, he rightly sensed that his theory would run into a wall of scientific criticism.

The study of 19th-century scientific views of the natural world, to say nothing of the history of the reception of Darwin's theory, constitute vast fields of scholarship that have generated a massive literature. In no way do I pretend to do them justice here. My purpose is simply to underscore the point that contrary to modern culture's assumption of inevitable support of evolution by science, Darwin's argument in its own day was an explicit counterpoint to an entrenched scientific position of long standing. This angle of vision allows us to see the actual historical context of *On the Origin of Species*, and provides a useful interpretive tool for understanding why Darwin characterized its text as "one long argument" (459).

A RELIGIOUS ODYSSEY

Of course, it was not only scientific proponents of special creation who found Darwin's theory objectionable. So too did many religious believers for whom God's direct creation of species was an important element of their faith. Radiating outward from its challenge to direct creation, the theory presented shocking challenges to other traditional formulations of Christian doctrine. Wherever the Genesis texts were read literally, it raised doubts about biblical authority. In addition, it removed the necessity for repeated miraculous interventions of a Creator, thus requiring new thinking about how God acts in the world. Given the magnitude of death and extinction that the theory requires, it increased the problem of suffering to staggering proportions. It introduced chance to the story of life, raising questions of providence and ultimate purpose. With regard to the human species, the idea that people evolved from a lower order of animals seemed an affront to human dignity, for if that were true, whence the gift of human reason and free will, the sense of moral responsibility, and the idea that human beings, male and female, are created in the image of God? From there, the challenges ramified out to require new interpretations of the narrative of Adam and Eve, original sin, the consequent necessity of the atonement of the cross, and a host of related doctrines.

Given the vexed history between evolutionary ideas and religion that continues to this day, it is important to emphasize at the outset that Charles Darwin was not a crusader against religion. Rather, he was a passionate lover of nature who asked questions of natural phenomena and found answers that remained at the level of natural causes. That these causes replaced direct divine causality made him cautious, at times anguished, and created tension with his wife. Still, in the end he did not shrink from arguing persuasively and boldly for what he had figured out.

Was he himself a religious man? Biographers have traced in his life a trajectory from youthful faith to deism to agnosticism, rightly claiming that, while as with any schema this is too simple, it does shed a certain light. Darwin's father was a freethinker, meaning he opposed a literal reading of the Bible and sought truth on the basis of logic and reason rather than the dogmatic statements of authority. From his mother, a practicing Unitarian like all the Wedgwoods, he acquired a sensibility that acknowledged a great Creator of the world, though not belief in a trinitarian God nor the divinity of Jesus Christ. At Cambridge Darwin's theological studies gave him broad familiarity with the traditional doctrine of high Anglicanism and, since membership in the state church was a necessary condition for graduation, familiarity with its rituals. He had a checkered if reasonably devout background.

The young Charles found the natural world created by God beautiful and full of wonder; close observation of it brought him intense joy. As he reported at the *Beagle*'s first stop at the Cape Verde Islands off the African coast to take on water: "The scene, as beheld through the hazy atmosphere of this climate, is one of great interest; if, indeed, a person, fresh from the sea, and who has just walked, for the first time, in a grove of cocoa-nut trees, can be a judge of any thing but his own happiness."[27] As he later recalled, this kind of personal experience of nature was intimately connected with a sense of God. Being outside in the natural world, interacting with it, filled his mind with feelings of "wonder, admiration, and devotion" directed to the One who created it, sublime beyond words.[28]

The *Beagle* journals are punctuated with this awareness. Consider his exclamation upon encountering the incredible layers of life in the Brazilian rain forest for the first time: "Twiners entwining twiners, tresses like hair – beautiful lepidoptera – Silence, hosannah."[29] It was not only palm trees and jungle butterflies that awakened his sense of the transcendent. Contrasted with the rolling green hills of southern England, the sheer rocky vastness of the Andes mountains touched this chord, as seen in his account of climbing a mountain pass in Chile:

> When we reached the crest and looked backwards, a glorious view was presented. The atmosphere so resplendently clear, the sky an intense blue, the profound valleys, the wild broken forms, the heaps of ruins piled up during the lapse of ages, the bright coloured rocks, contrasted with the quiet mountains of Snow, together produced a scene I never could have imagined. Neither plant nor bird, excepting a few condors wheeling around the higher pinnacles, distracted attention from the inanimate mass. I felt glad I was by myself, it was like watching a thunderstorm, or hearing in the full orchestra a chorus of the Messiah.[30]

A heightened aesthetic response to the beauties of nature leading to grateful responsiveness to its Creator comes to explicit expression when Darwin compares the rich life of the Brazilian rain forests to the bare desolation of the Tierra del Fuego, seeing them both as "temples filled with the varied production of the God of Nature." No one, he wrote, "can stand in these solitudes unmoved, and not feel that there is more in man than the mere breath of his body."[31] From the general, holistic tenor of his early reflections, it can be surmised that this voyager encountered God *in* nature rather than primarily deducing God's existence *from* it, as did natural theology.

In his study of the London years, Jonathan Hodge observes that as Darwin's idea of natural selection matured, he continued to see God as great, indeed, too great to keep intervening in the world to create new species in all their detailed structures and habits. What is so great about God causing elemental atoms to flash into flesh, creating a rhinoceros here or a succession of mollusks there, time after time? Those who think this way have a cramped imagination. Rather, is it not more in keeping with the Creator's greatness that general laws to bring about such species naturally be instituted, just as Newton's physical science had shown the motion of the planets to be subject to law and not due to miraculous divine adjustments? That being the case, Darwin maintained that the naturalist could not expect to understand the divine intention directly from particular structures of individual organisms or relationships among species. To attempt to read the divine mind this way is to forget how far above human knowledge the ways of God reside. Far better to live with a humble heart, allow that evolution is God's method of creation, and turn the mind to what it is in fact fitted to do: figure out the laws of nature.[32]

At the time of writing the *Origin*, Darwin's religious stance might plausibly be considered that of a deist, that is, one who rejects revelation as a source of knowledge but is unwilling to regard the laws of nature as accidental. Rather, they may well reflect a higher purpose behind the order of nature. His position seems to be that there is space for both a naturalistic science of the origin of species and for a Creator whose laws make the process possible. This becomes clear toward the end of the book where the summary introduces several theological tropes, the most impressive being the Creator's breathing life into a creature. Due to the circumstance that the same poison often affects plants and animals in a similar way, "I should infer from analogy that probably all the organic beings which have ever lived on this earth have descended from one primordial form, into which life was first breathed" (484). *Origin*'s famous last sentence repeats this image: "There is grandeur in this view of life, with its several powers, having been originally breathed into a few forms, or into one; ..." (490). In *Origin*'s second edition, anxious to pre-empt religious objections to his theory, Darwin added three words to the breathing metaphor: "by the Creator." Thus readers were presented with "one primordial form, into which life was first breathed by the Creator;" and the book ended with the grand vision of life with its powers "having been originally breathed by the Creator into a few forms" which continue to evolve. Darwin's consistent position at this point is that appeal to natural causes does not necessarily negate an originating action by a divine Creator.

Yet as he got older, Darwin grew progressively away from even this form of belief. The problems presented by his studies were certainly one factor. Besides inculcating a skeptical habit of mind, his work did raise difficult questions about traditional religious beliefs which theologians at the time were not quick to tackle. Then too, attempts by scientific proponents of special creation to undermine his theory either directly or by sleight of hand led him to protect it all the more by downplaying or omitting references to the deity. Darwin's eventual agnosticism, however, was not simply the result of his science. The trauma of the human condition played a large role. When his freethinking father died, he was forced to confront what he called the church's "damnable doctrine" about the fate of sinners, finding himself deeply disconcerted by the teaching that disbelievers such as his father suffer endless punishment after death. His moral repugnance toward the doctrine of hell showed itself in the fierce objection, "I can hardly see how anyone ought to wish Christianity to be true."[33] More powerful yet as a dissuasion to faith was the death of his beloved daughter Annie. The tragedy of her innocent suffering and his loss muted any attraction he might have had for a caring, beneficent God. The year following the publication of *Origin* he wrote to his friend and supporter, the Harvard botanist Asa Gray, "I had no intention to write atheistically ... But I own that I cannot see, as plainly as others do, and as I should wish to do, evidence of design and beneficence on all sides of us. There seems to me too much misery in the world." At the same time, he continued, he was not content to view "this wonderful universe and especially the nature of man, & to conclude that everything is the result of brute force. I am inclined to look at everything as resulting from designed laws, with the details, whether good or bad, left to the working out of what we may call chance."[34] In later years, sadly, he wrote that even the beauty of the natural world ceased any more to awaken his admiration. Worn down by life, the spirit of this great naturalist who had once breathed a silent "hosannah" when stepping into a rain forest seemed now in old age to be muted. Still, in private correspondence he declared that he had never been an atheist "in the sense of denying the existence of a God."[35]

This story of a soul ends in ambiguity. The point of tracing such a religious odyssey is twofold. First, it may simply be of interest to readers to know Darwin's own relation to religious belief. Second, it needs to be clear at the outset that the dialogue this book will engage in does not take the author himself as a model of faith, consulting him in a sort of anachronistic WWDD, what would Darwin do. Nor does what follow look into *Origin* for a Christian view of God, even though the idea of natural laws exercising a kind of secondary causality is indeed a point of light. Instead, what is relevant for our purposes is the way he could see.

THE BEHOLDER

Consider this incident. Studying the vegetation on a heath one day, Darwin noted that where fenced-in enclosures had been erected, multitudes of Scotch firs had sprung up. The rest was barren.

> But on looking closely between the stems of the heath, I found a multitude of seedlings and little trees, which had been perpetually browsed down by the cattle. In one square yard, at a point some hundred yards distant from one of the old clumps, I counted thirty-two little trees; and one of them, judging from the rings of growth, had during twenty-six years tried to raise its head above the stems of the heath, and had failed. (72)

Picture this investigator on his hands and knees out on the open heath, examining the ground which from above looked void of life. Imagine him, perhaps with a magnifying glass, peering down to count the rings on one little stem to discover that it was actually a fir tree that for over two dozen years had tried to grow but had been chewed back. Envision the satisfaction as he grasped the relationship between plant, animal, and the way the landscape appeared as it did in this place. The *Origin* is rich with these revelatory moments.

Charles Darwin loved the natural world. Gifted with remarkable powers of observation, he poured his attention on organisms large and small, captivated by how they looked, functioned, and interrelated with each other. Bugs, barnacles, birds, bromeliads: whatever he saw awakened wonder. This quality of fascinated delight spills over into his writing. At the outset in *Origin* he confesses without apology that the perfection of species' structure and adaptation "most justly excites our admiration," citing as an example "the woodpecker, with its feet, tail, beak, and tongue so admirably adapted to catch insects under the bark of trees" (3). His gaze also catches anomalies. There are woodpeckers who lived in a place with no trees! Geese living high on dry land who seldom alight on water although they have webbed feet! Nature astonished him in both its design and lack of design. Looking closely he discovered its ways with interest, and in passages of simple beauty registered surprise and awe. Catch this aside in his discussion of bees: "He must be a dull man who can examine the exquisite structure of a [honey]comb, so beautifully adapted to its end, without enthusiastic admiration" (224). The intensity of his observation registered the depth of his love for the world. I suggest that this quality of seeing the world with attentive and loving care is profoundly religious. In no way am I proposing that we project this interpretation onto Darwin as a person. His own religious odyssey which led him away from Christianity has its own integrity and is to be respected. But I am proposing that the sustained attention he lavished on the natural world models something of keen religious value to those who approach his work from the perspective of faith.

In the climactic move of an engaging lecture on the purpose of Catholic higher education, theologian Michael Himes made a point that is relevant here. In a world where the loving kindness of God is everywhere present but often overlooked, he said, the church's sacraments break through the fog and call attention to this reality. Using embodied things like bread, wine, oil, water, they name and celebrate grace for a moment, thereby allowing divine presence to gain a stronger foothold in our lives. By extension, certain events, persons, words, objects, or rituals can be considered a kind of sacrament analogous to the church's seven. They allow the grace of God, everywhere present, to break through in this or that instance. To illustrate, Himes works with the poem "Hurrahing in Harvest," penned by the

19th-century English Jesuit Gerard Manley Hopkins. In the opening lines summer is ending. Sheaves of grain stand bound after the harvest; autumn winds blow; wavy clouds drift over violet hills hung with mist. The poet is walking through this beauty when insight strikes:

> These things, these things were here and but the beholder
> Wanting;[36]

Considering this the single most beautiful statement of the Catholic sacramental principle, Himes suggests that education is (or can be) training in sacramental beholding. The purpose of higher education is to turn our students into beholders.

To underscore this point, Himes escorts his listeners to the early twentieth century, to a talk on asceticism given by the influential thinker Baron von Hügel to a Christian student's association at Oxford University. Who was the most striking example of asceticism in the nineteenth century just ended, von Hügel queried? Beyond doubt, he answered, Charles Darwin. With immense discipline and over a long period of time, he focused his keen, powerful intellect and astonishing energy on painstaking observation of nature, from the varieties of barnacles to the shape of pigeons' bills. With clarity and intensity he saw what was there. In the process he discovered one of the deep and powerful forces of the living world, changing our imagination forever. This is true asceticism, von Hügel exhorted. Himes agrees and riffs further. Higher education succeeds when our students learn to cease looking in the mirror long enough to look out the window at what is really there. It is a Christian conviction that in seeing reality they will discover grace, the Love that undergirds all that exists.[37]

Charles Darwin was a beholder.

He observed the smallest detail with interest, and recorded his scrutiny with affectionate care. Investigating not only individuals in isolation but organisms in relation to each other and to their physical conditions of life, he imaginatively set everything in motion. From land masses to flowers to birds to mammals, all are transient, marked with the scars of history. Under his gaze the tapestry of life comes alive, so to speak. Coming decades of observation are presaged in a description of his first descent from the *Beagle* for an on-shore walk in the Cape Verde islands. After recording "treading on Volcanic rocks, hearing the notes of unknown birds, & seeing new insects fluttering about still newer flowers," he registered a personal response: "It has been for me a glorious day, like giving to a blind man eyes."[38]

Meticulous experiments supported his arguments. Trying to figure out how fresh-water plants could arrive on distant islands over a salty ocean, Darwin surmised that birds, who waded in muddy ponds, could be the carriers because seeds would cling to their feet. Could this be tested? He wrote:

> I do not believe that botanists are aware how charged the mud of ponds is with seeds: I have tried several little experiments, but will here give only the most striking case: I took in February three table-spoonfuls of mud from three different points, beneath water, on the edge of a little pond; this mud when dry weighed only 6 3/4 ounces; I kept it covered up in my study for six months, pulling up and counting each plant as it grew; the plants were of many kinds, and were altogether 537 in number; and yet the viscid mud was all contained in a breakfast cup! Considering these facts, I think it would be an inexplicable circumstance if water-birds did not transport the seeds of fresh-water plants to vast distances, and if consequently the range of these plants was not very great. (386–7)

Of course he drew on experiments with birds' feet to complete the argument.

At times his gaze focused not outwardly toward the natural world but inwardly as he imaginatively brought together different bits of knowledge to glimpse larger patterns.

> What can be more curious than that the hand of a man formed for grasping, that of a mole for digging, the leg of the horse, the paddle of the porpoise, and the wing of the bat, should all be constructed on the same pattern, and should include the same bones, in the same relative positions? (434)

On the ordinary view of the separate creation of each being, we can only say that it has so pleased the Creator to construct each animal this way. But if we suppose an ancient progenitor had its limbs arranged this way, then all descendants inherited the pattern. The bones might be enveloped in thick membrane to form a paddle or a thin membrane to form a wing, or they may be lengthened or shortened for some profitable purpose, but there will be no tendency to alter the framework. Indeed, the same names can be given to the bones in widely different animals. What a "grand natural system," formed by descent with slow and slight successive modifications!

Darwin's passion for nature reveals not a cynical dismissal of spirit from the world, but a fascinated love for the living beings that are actually there. Using philosophical analyses of the way we see, Sallie McFague has developed a strong comparison between the arrogant eye and the loving eye.[39] The arrogant eye stares at another in a utilitarian, objectifying way that subdues and controls; the loving eye pays patient, careful attention to the particularity of the other in a non-sentimental, vitally interested way that reverences its reality. Measured by such a gauge, Darwin's gaze is an exercise of the loving eye. Having declared that we see beautiful adaptations everywhere and in every part of the organic world, he takes pains to describe with awe and empathy a multitude of such beauties: the sleek structure of a beetle which enables it to dive through the water; the shape of hummingbirds' bills in relation to the flowers they sip; the hinge of a bivalve shell; plumed seeds wafted by the gentlest breeze. Then his rich perceptions accumulate to reveal wider patterns at work over time. All point to a great natural force that has shaped the living world through an evolutionary process.

The first sentence of the first chapter of *On the Origin of Species* begins with the words, "When we look ..." (7). As a biologist Darwin's distinctive gift was to envisage the origin of all living beings in relationship to one another and to their environment over a deep sequence of time. *Origin* invites readers on an adventure of discovery to look upon the world in all its grandeur as he did. Readers' own powers of observation and imaginative visualizing are engaged step by step as the story of life unfolds. In the next chapter we will take this journey.

3

"Endless Forms Most Beautiful"

> There is grandeur in this view of life, with its several powers,
> having been originally breathed into a few forms or into one; and
> that, whilst this planet has gone cycling on according to the fixed
> law of gravity, from so simple a beginning endless forms most
> beautiful and most wonderful have been, and are being, evolved.
>
> Charles Darwin

Unlike what a reader might expect, *On the Origin of Species* does not actually recount the story of life's beginnings on earth, nor does it trace in chronological order the sequence of species' appearing from earliest forms to the higher mammals, as if it were a history of evolution. In this sense it is not a story of origins at all. Instead, the book is structured as what the author calls "one long argument" (459), crafted to demonstrated that there is such a history to begin with. Its different chapters taken together make the case that over millions and millions of years species of plants and animals have descended from original parents, along the way diversifying and going extinct, due to the working of natural selection. In a word, the origin of species is from one another.

STARTING WITH FARM AND GARDEN

In a shrewd move the author begins like a good teacher whose pedagogical method includes starting with an apperceptive basis, with what students already know in order to move them into the unknown. Entitled "Variation under Domestication," Chapter 1 describes the familiar human practice of choosing certain animals and plants which have desirable traits and mating them in the hope of reproducing and strengthening these features. In England at the time, people bred cattle for choice types of meat, cows for udders that yielded more milk, race horses for speed, grains for yield, apples for taste, hyacinths and dahlias for color, size, and scent, fancy pigeons for pleasure. Many of *Origin*'s readers possessed direct experience of how a woolier sheep or a faster greyhound could be produced. Indeed, breeders nurtured a host of agricultural, culinary, orchard, and garden species into new forms with characteristics that were useful at different seasons. Domestic breeding shows that variations can be inherited, *Origin* points out, and that the steady accumulation of a trait over successive generations can effect dramatic change.

In order for breeding to succeed, three conditions must be met. To begin with, there must be variations in a species, so people have qualities from which to choose. Next, there must be an act of selection whereby breeders decide to enhance a certain property. Finally, the property must be able to be inherited. If offspring show up with the desired trait and are then bred with another similarly endowed, the trait will begin to establish itself in the population. Over time, a breed will exhibit the desired quality with increasing frequency. Advantageous variations, selection, and inheritance: with these components in action, people could modify the properties of domesticated plants and animals.

As an example of wildly successful breeding, Darwin appealed to pigeons. Cultivating this bird was an immensely popular hobby of many of his fellow citizens. "Believing that it is always best to study some special group, I have, after deliberation, taken up domestic pigeons" (20). He built a shed at Down House and kept every breed which he could purchase or obtain for live study and anatomical dissection; at one point his flock grew to 90 birds. Diplomats in India and Persia send him skins of native pigeons. He read treatises, exchanged letters, associated with several eminent pigeon fanciers, and joined two London pigeon clubs. The verdict: "The diversity of the breeds is something astonishing" (21). In a dazzling nine-page riff on their variations, he carefully compares the "wonderful differences" found in the trumpeters, carriers, tumblers, pouters, runts, fantails, Jacobins, laughers, barbs, dragons, and other breeds. Virtually every anatomical detail differs: beaks, eyelids, nostrils, and mouths; necks, wattles, wings, tails, and feet; color, shape, size, and placement of feathers; skin wrinkles on head and toes; skeletal skulls and jaws; the shape, size, and thickness of their eggs; their manner of flight and the sound of their voices. Breeders competed to produce varieties of pigeons with a prize-winning color or beak shape. Darwin's observations were in service of showing how much variation could exist within a single type of animal.

A deeper question accompanied these descriptions: where did all these different breeds of pigeons originate? Breeders for the most part thought that each type of pigeon descended from its own particular ancestor, so that the current stock mirrored a pre-existing range of original wild stock. In a series of pragmatic arguments Darwin shows how this is not likely to be the case. Despite their distinctive looks, when crossbred with one another they produce fertile offspring. This indicates that they all descend from the same aboriginal form. "Great as the differences are between the breeds of pigeons, I am fully convinced that the common opinion of naturalists is correct, namely, that all have descended from the rock-pigeon (*Columba livia*)" (23), a blue-gray bird with black bars on its wings. The importance of this move already in the first chapter cannot be underestimated. If a blooming variety of pigeons could descend from one ancestral species as a result of domestic breeding, then the effects of selection can be far-reaching.

As to how desired variations actually originate, in the mid-19th century Darwin was aware of the dearth of knowledge on the subject. "The laws governing inheritance are quite unknown" (13), he wrote with some frustration. Vexation with this ignorance will be expressed numerous times in the coming chapters. He rightly suspects, however, "that the most frequent case of variability may be attributed to the male and female reproductive elements having been affected prior to the act of conception" (8). Environmental factors may also play a small role, as may the use or disuse of a property. Wherever it comes from, however, once a variation shows up and is thought to be advantageous, human beings can work with it, selecting and perpetuating it into the next generation. "The key is man's power of accumulative selection: nature gives successive variations; man adds them up in certain directions useful to him" (30). This requires sharp powers of observation, since slight variations are not immediately obvious.

Skillful breeders with accuracy of eye and judgment are one in a thousand. Still, when attention is paid and the right choices made, "the great power of the principle of selection is not hypothetical" (30). New types of animals and plants with preferred properties result.

Darwin's own dining table benefitted from this process with the strawberry. No doubt this plant had always varied, but the slight variations had been neglected. "As soon, however, as gardeners picked out individual plants with slightly larger, earlier, or better fruit, and raised seedlings from them, and again picked out the best seedlings and bred from them, then there appears (aided by some crossing with distinct species) those many admirable varieties of the strawberry which have been raised during the last thirty or forty years" (41–2).

With these and numerous other examples of the human practice of choosing advantageous characteristics in plants and animals, Darwin seeks to persuade readers of two truths. First, species are not set in stone but are "plastic" (12), able across generations to depart from the parent type to yield vastly different varieties and sub-varieties. Second, selection is the means by which these good results are obtained. Selection changes traits. Its importance consists "in the great effect produced by the accumulation in one direction, during successive generations, of differences absolutely inappreciable by the uneducated eye – differences which I for one have vainly tried to appreciate" (32). This author may not have had the talent to become an eminent breeder, but he identified the principle, namely the cumulative power of selection, by which good breeders succeed.

None of this is especially controversial. By starting with domestic breeding practices Darwin astutely sets the stage for the next step in the argument of *Origin*, which uses the human selection of domesticated species in farms and gardens as an analogy for a similar selective process going on in undomesticated nature. In the coming pages the analogy will be explicit:

> I have called this principle, by which each slight variation, if useful, is preserved, by the term of Natural Selection, in order to mark its relation to man's power of selection. We have seen that man by selection can certainly produce great results, and can adapt organic beings to his own uses, through the accumulation of slight but useful variations, given to

him by the hand of Nature. But Natural Selection, as we shall hereafter see, is a power incessantly ready for action, and is as immeasurably superior to man's feeble efforts, as the works of Nature are to those of Art. (61)

TWO KEY ELEMENTS: VARIATION AND STRUGGLE

As with human practices of selection, nature's work requires first of all the existence of desirable qualities that can be selected, and these need to be inherited. With a multitude of observations and experiments, Chapter 2, entitled "Variation under Nature," advances the argument that such is the case. Nature is filled with a profusion of related types of organisms which vary in small but significant ways from one another. Even among offspring of the same parents one can find individual differences. Every cow and dog, every tree and ear of corn gives evidence of variability; no two are identical. Variations do exist in nature; they have pronounced effects on the fitness of organisms to live and reproduce; they can be passed on to the next generation.

The prevailing idea at this time was that each species was separately created and its characteristics remained fixed over time. Consequently, it was assumed to be relatively easy to identify organisms as belonging to one species or another. In the eighteenth century the Swedish naturalist Carl Linnaeus had devised a system of classifying plants and animals which proved to be a tool of considerable utility; with significant revisions it is used to this day. At the very broadest scale all natural things are assigned to one of three kingdoms, animal, vegetable, or mineral. Kingdoms are divided into phyla and thence into classes; classes split into orders; orders branch down into families; families ramify into genera (singular: genus); and each genus is comprised of species. A wolf, for example, belongs to the species *lupus* of the genus *Canis*; continuing up the ladder of classification its genus fits into the family *Canidae*, the order *Carnivora*, the class *Mammalia*, the phylum *Chordata*, and the kingdom *Animalia*. Within species there are multiple sub-species or varieties. Once enough organisms had been categorized in this system, the world could be approached in a clearly organized way. In Linnaeus' view, his classification reveals the very plan of creation. In his own inimitable words, "God created, Linnaeus arranged."[1]

Darwin's emphasis on variation pulls counter to the neatness of this system. Tongue in cheek, the author noted that "It should be remembered that systematists are far from pleased at finding variability in important characters ..." (45). With a keen eye for what didn't fit, he advocated the importance of slight differences, intermediary forms, weak variations that seemed to get stronger over time. It pleased him to note that 182 British plants considered varieties by some botanists were each also ranked as species by others. Part of the problem, Darwin argues, lies with the standard idea that species are stable, immutable entities. Many years ago when he and others set about comparing birds from the separate islands of the Galápagos Archipelago, both with one another and with those from the American mainland, he was struck with how vague and arbitrary is the distinction between species and varieties. Until an agreed-upon definition of these terms could be reached, arguing over whether a group was a species or a variety "is vainly to beat the air" (49).

Rather than classifying organisms according to their distinct forms, which naturalists deduce from their external characteristics and philosophers understand to be based on an archetype or fixed essence in a metaphysical sense, Darwin urged that it is better to think of nature in the following way. Individual differences are first steps toward slight varieties, which blend into well-marked varieties, which merge into sub-species, which become distinct species, which form large genera, the whole forming an insensible series. This series "impresses the mind with the idea of an actual passage" (51). Species once existed as varieties and have so originated; varieties are incipient species. An evolving historical relationship between organisms offers a distinctly different approach to classification. An organism should not be fixed in a category like a specimen butterfly on a corkboard, but traced in its movement within the larger story of life's actual passage. No longer a collection of separate entities, the forms of life throughout the world become divided into groups emerging from groups along a beautiful narrative arc.

It is clear that far from simply establishing the existence of inherited variations in nature, this second chapter uses the theory of evolution to interpret the puzzling range of data that variations present. There is a certain flexibility in the constitution of species. Natural selection works on numerous small differences in a gradual but cumulative fashion so that changes amplify over time, and new species emerge.

But now a new question arises. In breeding domesticated species, human beings deliberately choose properties that are desirable to themselves. Given that nature does not act in a similarly conscious way, what criterion governs the selection? What makes nature select the way it does? To answer this question, Darwin paints a picture of the circumstances in which selection occurs. The element of struggle becomes the next key feature of the theory, laid out in Chapter 3 entitled "Struggle for Existence."

During his London years Darwin had read Thomas Malthus' influential *Essay on the Principle of Population*. Its thesis held that human populations tended to grow at a geometric rate, faster than their ability to produce food which increased only at a mere arithmetic rate. The resulting gap between human need and available resources would create inevitable clashes as people vied for what sustained life. Over the next forty years Darwin found this construal useful in the development of his theory about the natural world. All organic beings tend to increase at a high rate. In every species more individuals are born than can possibly survive, given the ultimately limited resources of the Earth. A strenuous effort to access these assets is thus inevitable. In the process, some destruction of life occurs; without it there literally would be no standing room left. In this circumstance, nature applies a system of checks and balances. We see species eating but also serving as prey, spreading out but being chewed back, fighting off or succumbing to disease, surviving or dying in severe weather events, waging battle or setting up novel forms of cooperation, with varying success.

Two dogs in a time of dearth clash over which shall get food and live. Mistletoe growing on an apple tree endeavors to be more attractive than other fruit-bearing plants so that birds will devour its seeds and its offspring will spread. A plant on the edge of the desert, dependent on moisture, fights for life against the drought. A tree which annually produces a thousand seeds competes with other trees over ground space where even one seed can take root. And so on. "In these several senses, which pass into each other, I use for convenience sake the general term of struggle for existence" (63).

Like natural selection, the term is a metaphor: "I use the term Struggle for Existence in a large and metaphorical sense, including dependence of one being on another, and including (which is more important) not only the life of the individual, but success in leaving progeny" (62). The word *struggle* expresses the interpenetration of energies. Note from the outset that it includes not just competition with others but also mutual dependence or cooperation and achievement of producing offspring. It refers to the whole dynamic range of relationships—with other members of the same species, with the next generation, with individuals of other species, and with the physical conditions of the environment—that shape an organism's effort to survive and reproduce. This is the context for natural selection's choice:

> Owing to this struggle for life, any variation, however slight and from whatever cause proceeding, if it be in any degree profitable to an individual of any species, in its infinitely complex relations to other organic beings and to external nature, will tend to the preservation of that individual, and will generally be inherited by its offspring. (61)

To an audience steeped in natural theology's worldview, nature was largely a harmonious place, operating according to a pre-set design. Many paid no heed to the real state of nature in which survival is not assured. To be realistic, it is necessary if difficult to acquire an unsentimental understanding of "the mutual relations of all organic beings." In a lyrical passage *Origin* lays out what this entails:

> We behold the face of nature bright with gladness, we often see superabundance of food; we do not see, or we forget, that the birds which are idly singing around us mostly live on insects or seeds, and are thus constantly destroying life; or we forget how largely these songsters, or their eggs, or their nestlings, are destroyed by birds and beasts of prey; we do not always bear in mind that though food may be now superabundant, it is not so at all seasons of each recurring year. (62)

In a finite environment, plants and animals have to find water, food, space to live, and ways to reproduce or they and their kind will not survive. Thus they contend with each other, or in some instances cooperate, in constant response to life's deep imperative. A remarkable aside offers comfort to the reader in the face of so much death and destruction: "When we reflect on this struggle, we may console ourselves with the full belief, that the war of nature is not incessant, that no fear is felt, that death is generally prompt, and that the vigorous, the healthy, and the happy survive and multiply" (79). Perhaps.

Darwin had a knack for seeing informative patterns in seemingly mundane places, and illuminates the checks and balances of nature with interesting examples. In one experiment he dug up and cleared a plot of ground three feet wide by two feet so that it was free of rooted plants. "I marked all the seedlings of our native weeds as they came up, and out of the 357 no less than 295 were destroyed, chiefly by slugs and insects" (67). Another time he cordoned off a plot already growing with natural vegetation. As the season progressed it was not slugs but the more vigorous plants that wiped out those less adapted: "Thus out of twenty species growing on a little plot of turf (three by four) nine species perished from the other species being allowed to grow up freely" (68).

A colorful instance of animals and plants bound together by "webs of complex relations" can be found around English villages, where a profusion of flowers can be found as compared with the more distant countryside. Certain flowers require fertilization by humble-bees (the common term for bumblebees). The number of these insects in any district depends on the number of field mice, which destroy their honeycombs and nests. The number of mice, in turn, is largely determined by the number of cats in the vicinity which find them irresistible morsels. Hence near villages where people keep cats, "it is quite credible that the presence of a feline animal in great numbers ... might determine, through the intervention first of mice and then of bees, the frequency of certain flowers in that district" (74). The more cats, the fewer mice, the more bees, and the more flowers due to complex biotic interactions.

Further afield, a forest in the southern United States provides an international example. Centuries before colonization the native inhabitants cut down part of an original forest to clear living space. Once the Indians abandoned the site, a succession of shrubs, bushes, and smaller trees moved in, each replacing the other until something like the original configuration of species was restored. Now the trees growing on the ancient Indian mounds display the same beautiful diversity and proportion of kind that is seen in the surrounding uncut forests.

> What a struggle between the several kinds of trees must here have gone on during long centuries, each annually scattering its seeds by the thousand; what war between insect and insect—between insects, snails, and other animals with birds and beasts of prey—all striving to increase, and all feeding on each other or on the trees or on their seedlings, or on

the other plants which first clothed the ground and thus checked the growth of the trees. (74–5)

The long interplay between countless plants and animals has determined the face of the forest that so attracts us today.

In this endless interaction, certain variations show up which give their owners a small advantage. They give organisms an edge in the search for nutrition and successful reproduction. Such variations spell success and, as they get passed on, become adaptations that spread through the group. Even the most trivial characteristic, such as the down on a fruit's skin or the color of its flesh, might be acted on by natural selection if such would repel insect attacks. The tail of the giraffe, used for swatting flies, might seem too trifling a trait to be adapted by successive slight modifications. Yet Darwin drew on observations from his *Beagle* voyage to make the case for the tail's importance. The distribution of cattle in South America, he recalled, absolutely depends on their power of resisting the attacks of insects. It is not that the quadrupeds are actually destroyed by the flies, but they are incessantly harassed and their strength reduced, so that they are more subject to disease and not as able to escape from larger predators. Individuals which could defend themselves from the hounding of these small flying pests would gain a great advantage, being stronger and able to range into new pastures. The same would be true anywhere flies torment larger beasts. Thinking of a different species on a different continent, Darwin concludes that the tail of the giraffe in Africa would be selected for this reason.

Origin presses the argument about this dance of life with multitudes of examples from observations of wheat, sweet-peas, sheep, swallows, rats, cockroaches, and many more. The innumerable interactions of plants and animals, their effort to survive during long centuries, determine the proportional numbers and kinds of creatures living in any country (in our terms, ecosystem) at any given time. What results from these ongoing, multidimensional, biological interactions are the landscapes, the beautiful entangled banks, that so enchant us.

This is a deeply ecological vision of nature. It entails a network of intricate interdependencies and mutual relations expressed in competition or profitable cooperation. Over time, each being's structure becomes related, in the most essential yet often hidden manner, to that of all other organic beings with which it competes for food, or on which it preys, or from which it has to escape, or near which it has to reproduce, or with which it cooperates. Think of the teeth and talons of the tiger, or the hook of the parasite which clings to the tiger's fur. Every variation which gives its owner a certain advantage in the struggle for existence stands in the closest relationship with the land, climate, and other creatures in the area. As these are preserved, accumulated, and inherited, they slowly and beautifully adapt each form to the environment in which it lives.

Is this process haphazard? When we look at a beautiful entangled bank without reference to a divine designer, we might be tempted to impute its pleasing proportions, its numbers and kinds of species, to what we call chance. Surprisingly, Darwin rejects this idea: "But how false a view this is!" (74). The natural world's beauty is due to the mutual interactions of species in the struggle for life. It is due to birth, thriving, suffering, and death which have brought forth organic beings having beneficial dependencies upon each other or advantageous adaptations over their competitors or enemies. There is a sense to it, a reasonable explanation for it. Mutual interplay has created the living world as we know it.

THE THEORY: NATURAL SELECTION

Up to this point Darwin has been laying the groundwork for his theory about how species originate. Notice how human beings breed domestic animals and plants by selecting for useful traits; the same can be done in wild nature. Consider how variations occur in nature; these can be inherited. Observe how an ongoing struggle for existence plays out in the environment; some variations give an advantage to their owners. *Origin* now pulls these key elements together in a sustained presentation of Darwin's major insight, climaxing in the metaphor of the tree of life. The driving force that works on variations in a situation of struggle to bring about the origin of species is, as Chapter 4's title announces, "Natural Selection."

Darwin reiterates: In the course of thousands of generations, conditions of life that most likely affect the reproductive organs give rise to variations. Some of these give individuals a slight advantage to survive and produce offspring, thus passing on the adaptation. Others adversely affect the organism, and these are destroyed by virtue of the individual's not surviving or reproducing. "This preservation of favourable variations and rejection of injurious variations, I call Natural Selection" (81). Unless variations occur, natural selection can do nothing. But once they arise, this principle springs into action, acting analogously like human breeders in choosing to promote a desirable trait. It acts not only on visible characteristics, as human breeders do, but "on every organ, every shade of constitutional difference, on the whole machinery of life," (83) seeking advantage or disadvantage that may well tip the precariously balanced scale in the struggle for life, and selecting for the advantage.

In the short run, individuals are preserved. In the long run, their favored bodily structure and abilities are perpetuated, spreading throughout the population. The outcome is organisms ever more beautifully adapted to their life's situation. Leaf-eating insects are green, bark-feeders mottled grey, the alpine ptarmigan white in winter, the red grouse the color of heather, and the black grouse that of peaty earth. Why? These tints serve to preserve them from dangerous predators, and so are chosen by natural selection. Flowers excrete nectar which attracts insects needed to pollinate the plants. Favor is to the flowers which excrete the most desirable or well-located nectar, thus dusting insects with their pollen, thereby producing the most vigorous seedlings via the best adapted bugs, also selected for their sipping apparatus that enables them to get at the sweet juice. Seemingly mundane interactions like the transfer of pollen by insects can explain the structure of flowers which so enchant us with their beauty; the complex interaction between plant and pollinator drives both floral and insect evolution.[2]

Under the pressure of selection, small advantageous differences increase steadily, resulting in breeds that ultimately differ in character from each other and from their common parent. Wolves hunt for prey; the fleetest have an advantage in capturing food over slower and heavier members of the pack; selection allows descendants of the swift to increase, and thus over thousands of generations the trait spreads slowly through the population. Over whole geological periods and thousands of generations, the result is the origin of new species and interacting communities, united in mutual relations that are infinitely complex and close-fitting.

Geology provides an important analogy. Before Charles Lyell's work, people thought coastal waves were a trifling and insignificant cause when it came to carving out gigantic valleys or to forming long lines of inland cliffs. But now one seldom hears objections to the idea that such changes of the earth were formed gradually. So, too, "as modern geology has almost banished such views as the excavation of a great valley by a single diluvial wave, so will natural selection, if it be a true principle, banish the belief of the continued creation of new organic beings, or of any great and sudden modification in their structure" (95–6).

Two dynamic principles amplify the outcome of natural selection, acting like its right and left hands, namely, divergence and extinction. The significant principle of divergence refers to the splitting of species into new varieties and species, rather than one species simply morphing straight as an arrow into another. It is premised on the idea that more life can flourish in an area if it is occupied by different types of organisms that draw from the same resources in different ways. If a region is filled to capacity with a species of carnivorous quadruped, for example, its numerous descendants will thrive only if they diverge to feed on new kinds of prey, climb trees, take to the water, or become less carnivorous. A plant that multiplies in a restricted area will flourish only if some of its members develop deeper roots to find water or a taller stem so branches can catch moisture from the air. Since the greatest amount of life in one place can be supported by the greatest diversification of structure, the pressure of selection will favor animals and plants that can seize on unexploited places and roles in the natural economy. They will diverge, becoming more and more dissimilar over time, their descendants competing less directly. Species will trend toward splitting and spreading.

In addition to branching out in one habitat, divergence likewise explains the pressure for organisms to occupy living places at a distance, bringing the species into new conditions that may trigger yet more variations. Ducks and hawks both descend from one ancestral bird species. Today ducks are fitted to diving in water for their food, while hawks are adapted to swooping through air. Divergence caused their increased specialization in relation to the surrounding environment. "The more diversified the descendants from any one species become in structure, constitution, and habits, by so much will they be better enabled to seize on many and widely diversified places in the polity of nature, and so be enabled to increase in numbers" (112).

The principle of extinction also looms large in the working of natural selection. As selected and favored forms increase in number, filling niches and consuming resources, so will the less favored forms decrease and become rare. Ancestor species and transitional forms in particular are likely to diminish as their better adapted progeny multiply. Reduced numbers indicate that they are on the way out: "Rarity, as geology tells us, is the precursor to extinction" (109). Small and broken groups and sub-groups will tend to disappear. As a result, "many ancient forms of life have been utterly lost" (431). Which groups will ultimately prevail and which vanish, no one can predict. But many once thriving species are now extinct while better adapted ones have taken their place. Extinction due to failure to adapt to changing conditions is inevitable, "for the number of places in the polity of nature is not indefinitely great" (109).

Along the main line of evolution divergence drives lineages apart and extinction erases evidence of the transition. As these forces shape species, Darwin sees another dynamic which he calls sexual selection operating alongside. Sexual reproduction is preferable to self-fertilization or inbreeding, providing for combinations that produce new and vigorous varieties. In that context, sexual selection goes into gear. The context here is not the struggle for existence but a struggle between males for possession of the females. The female will choose the healthiest, most attractive partner. Consequently males engage in behavior that seeks to attract female interest or intimidate rivals: aggressive shows of strength, special vocalizing, careful nestbuilding, or colorful displays like the outsized antlers of the Irish elk or glorious plumage of the peacock. Such display of their attributes as good reproducers may actually be maladaptive, large antlers hindering easy movement through the forest or the peacock's tail preventing easy flight. But it also gains sexual partners. While the outcome for the unsuccessful competitor in reproductive competition is not death, it does leave him without offspring. Over time this process of sexual selection can produce a marked effect on a species.

Since natural selection can act only by the preservation and accumulation of infinitesimally small modifications, each profitable to the preserved being, Darwin envisioned it as a slow, intermittent process. Circumstances favorable to its working are vast periods of time, large spaces, and great numbers of varying individuals and species. Then beneficial variations, at first barely appreciable, steadily increase, and new breeds emerge that trend away in character both from each other and from their common parent.

THE TREE OF LIFE

To enable the reader to fathom how natural selection with its auxiliary principles of divergence and extinction works to bring about the world we see today, Darwin drew a diagram of taxa, or units of natural life. Instead of picturing concretely the evolution of finches, fishes, or frogs, it is a schematic device using letters and numbers that aims to represent abstractly how the dynamic process of evolution leads to a divergence of forms over millions of years. While the series of dotted lines hardly indicates a beautiful picture of trees or gulls, if the reader's imagination would follow the diagram as if it were a slow motion film, the magnitude of what he is proposing becomes clear.[3]

Let A to L, he begins, represent different species of a large and vibrant genus.

Recall: genus and species are categories of biological classification. Although there are no hard and fast definitions, genus is the more inclusive category. It is comprised of a group of similar species that resemble each other but cannot successfully interbreed to produce fertile offspring. By contrast, "A species is a population whose members are able to inter-breed freely under natural conditions."[4] The genus *Panthera* (big cats), for example, is comprised of four species: lion, tiger, jaguar, and leopard. In

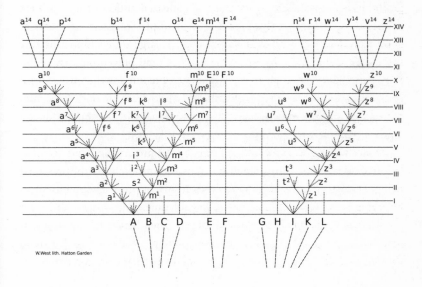

W. West lith. Hatton Garden

the wild, lions and tigers do not mate; even if they did and were able to produce cubs, the offspring would be sterile. *Pinus* (pine tree) is a genus comprised of trees which resemble each other insofar as they all have long, narrow needles bound in bundles and hard, woody cones with thick, tough scales. There are about 115 species of pine trees worldwide, including the Scotch pine, sugar pine, ponderosa pine, lodgepole pine, Turkish pine, etc. Being separate species, however, they cannot fertilize each other's pine cones. As of this writing, biologists have classified about 1.7 million species of plants and animals currently alive on the planet. This is estimated to be less than one-quarter of the total of living species, the remainder comprised mostly of bugs and bacteria. The number is diminishing rapidly due to the current wave of extinctions; already gone in the twenty-first century are species such as the Baiji dolphin, the West African black rhino, the golden Monteverde toad, and the Hawaiian crow.

Let A to L, Darwin declares, represent different species of a large and vibrant genus. Let A be a common, widely diffused, successful species. The little fan of diverging dotted lines proceeding upward from A represents the variations that show up from time to time in the original stock. Only those variations which are in some way profitable will be naturally selected; the rest will be rejected. The little dotted sprays that peter out represent unsuccessful branches of descendants that then disappear. The dotted lines that reach a horizontal line and are given a lower case letter represent accumulated changes that are so beneficial they stand out as new, markable varieties.

The intervals between the horizontal lines in the diagram may represent a thousand generations, or a million or even a hundred million generations. Let us stay with a thousand. The chart supposes that after the first thousand generations, species A will have produced two strong varieties, a^1 and m^1. These will tend to inherit the advantages which made their common parent numerous and productive. During the next thousand generations they will go on bringing forth variations, some of which will be preserved by natural selection, the rest of which will not successfully breed. After this interval of another thousand generations, a^1 will have produced variety a^2 which, owing to the principle of divergence, differs more from parent A than did variety a^1. Meanwhile, m^1 will have produced two varieties, m^2 and s^2, which differ considerably from each other and even more so from their common parent.

The process continues by similar steps for unthinkable lengths of time. The modified descendants proceeding from the common parent A will generally go on increasing in number and diverging in character with more or less success. So a^3 generates a^4 which continues the line of succession, but also diverges to varieties d^4 and d^5 which eventually disappear. Variety a^4 produces a^5, which itself continues reproducing to a^6 but also generates a newly successful variety, f^6. The diagram tracks the course of life up to the fourteen thousandth generation.

Obviously, the process is not so neat. Exceedingly complex relations are in play including other nearby species that are developing, empty niches that open up, and climate change, among other factors. But as a general rule, the more diversified in structure the descendants of the parent species become, the greater their advantage, and the more their modified progeny will multiply and produce new distinct variations.

After ten thousand generations, species A will have produced three viable forms of life, a^{10}, f^{10}, and m^{10}. Their history of divergence over successive generations means they will differ greatly from each other and from their common parent. It may even be the case that by this time they have become well-defined species in their own right, no longer able to breed with each other. "Thus the diagram illustrates the steps by which the small differences distinguishing varieties are increased into the larger differences distinguishing species" (76).

Condensing the process, the diagram's schema of the next 4,000 generations (or four million or 400 million generations) results in eight different species, numbered a^{14} to m^{14}, all descended from A. Here is a graphic illustration of the principle expressed so clearly by Alfred Russell Wallace: Every species has come into existence coincident both in time and space with a pre-existing closely allied species.

By similar steps, species I, a second starting parent, has after fourteen thousand generations produced six new species, labeled n^{14} to z^{14}. An unusual case is found in species F. Only F may be supposed to have transmitted offspring (F^{14}) to the fourteen thousandth generation unchanged or altered only in a slight degree—such would be the horseshoe crab or the gingko tree in our era.

During this whole process of descent with modification, extinction plays an essential role. With natural selection favoring whatever advantage they have over other forms in the struggle for life, the improved descendants of any one species will constantly tend to branch out, supplant, and so ultimately destroy earlier varieties in the line of generation. Original forms will be squeezed out by the modified ones, which occupy an increasing diversity of niches or habitats. These rising species will likely cause not only their immediate predecessors but also their original parent to go extinct, unless these adapt or find a niche where they can thrive. In the ordinary course of events, by the time we reach the fourteen thousandth generation parent-species A and all the intermediate varieties will have been lost, replaced by eight new species. Parent-species I will likewise be extinct, having given way to six new species. Though the other nine species of the original genus, marked B to L, may for a long time continue to transmit unaltered descendants, most also go extinct, except for F. This is shown in the diagram by dotted lines prolonged upward which then peter out. Better adapted to the natural environment, the species at the top of the diagram have supplanted all but one of the starting species.

Note that from the original eleven species, the diagram ends up with fifteen: life has diversified. Eight of the original eleven species have no descendants at all: they have gone extinct. One species has descendants that have barely changed at all: these appear archaic, like living fossils, in the new era. All the other new species come from two of the original eleven. Owing to the divergent tendency of natural selection, there will likely be much greater differences among these last species than between the original eleven. These new species, moreover, will be related to each other in a widely different manner. Of the eight descendants from A, the three on the left will be closely related due to having recently branched off from a^{10}. Since they forked off from a^5 at an earlier period, the two new species in the center will be distinctly different from those three. Lastly, from having diverged much earlier in the process of modification, the three on the right (o^{14}, e^{14}, and m^{14}) will be closely related to each other but will differ widely from the other five species. Over time each of the three groupings may well even constitute a new genus. Given that the same is true of the descendants of I, the diagram charts how multiple new genera may be produced by descent with modification from two species of one older genus.

One final point: in the diagram, the broken lines beneath the capital letters of the original eleven species converge downward toward a single point. This point represents one species of an even earlier, ancient and unknown progenitor from which the parent species A and I themselves descended. The process has gone on for as long as life has existed on Earth, back to what is today called a universal common ancestor.

In a way more graphic than words, this diagram charts the cumulative outcome of the births, lives, and deaths of a few species over an unimaginable run of times past. Say this diagram simply charts the evolution of beetles or of orchids. Expand this pattern of repeated forking and branching, adaptation and extinction, to every creature alive at the same time, all interacting under the pressure of selection in the struggle for existence. It boggles the mind:

> Whatever the cause may be of each slight difference in the offspring from their parents and a cause for each must exist, it is the steady accumulation, through natural selection, of such differences, when beneficial to the individual, that gives rise to all the more important modifications of structure, by which the innumerable beings on the face of this earth are enabled to struggle with each other, and the best adapted to survive. (170)

During the generations of descent with modification, it is impossible to predict which particular groups will prevail. Whichever ones produce adaptations that give them an advantage in view of the actions of other species and the environmental conditions of life at any given time, these will survive and reproduce. The rest will fade from view. "Thus, as I believe, species are multiplied and genera are formed" (120).

Recall the standard idea in nineteenth-century science that all species are immutable, each being exquisitely designed and located in its time and place by a special act of the Creator. Operating with that assumption, Linnaeus' work of classification identified organic beings based on visual similarities and differences. In face of that paradigm, the theory of evolution argues that the true basis of classification is genealogical. A community of descent is the hidden bond that ties together all living beings into one narrative of life and death stretching over millions of years. Darwin's theory uncovers the inner affinity of all organic beings to one another, rather than their merely external relations. This realization strikes him with awe:

> It is a truly wonderful fact – the wonder of which we are apt to overlook
> from familiarity – that all animals and all plants throughout all time and
> space should be related to each other in group subordinate to group, in
> the manner which we everywhere behold. (128)

Darwin devotes ten pages to discussion of this diagram in Chapter 4, and twenty-five pages more in Chapter 13. The only drawing in *Origin*, it is a profoundly influential interpretive device that allows scientists and general readers alike to get an imaginative grasp of the working of evolution. Illustrating at a glance how the entire history of life can be seen as a phenomenon of gradual modification, splitting, divergence, and extinction of species, it "constitutes one of the most spectacular examples of a shift of paradigm"[5] to this day.

In Darwin's fertile imagination, this diagram converts, finally, into the metaphor of the tree of life. Picture a spreading evolutionary tree that links nature and history into an indivisible whole, spanning the ages. The outer layer of budding twigs and green leaves represents the multitudes of species alive today, topping out in the sun. They connect to major limbs branching out from the trunk, representing previous periods of growth of the main groups of organisms; these limbs were themselves once budding twigs when the tree was small. They have forked into smaller limbs which divide into lesser and lesser branches, signifying the splitting into multiple descendant species. The flow of life is created by infinitesimal gradations, since "in a tree we can specify this or that branch, though at the fork the two unite and blend together" (432). Lower down are dead branches, standing for the long succession of extinct ancestral forms. Only two or three twigs which flourished when the tree was a mere bush may have survived as long shoots, as is the case with very few ancient species which still have living descendants. Many a limb has decayed and dropped off, representing whole families of organisms which are known only from the fossil record. Over millions upon millions of generations, all living beings are connected in this grand flow of life:

> As buds give rise by growth to fresh buds, and these, if vigorous, branch
> out and overtop on all sides many a feebler branch, so by generation I
> believe it has been with the great Tree of Life, which fills with its dead
> and broken branches the crust of the earth, and covers the surface with
> its ever branching and beautiful ramifications. (130)[6]

Over incredibly long ages and diverse conditions organic beings have produced variations; some of these are useful in the struggle for existence; nature selects for these advantages; selected organisms diverge into new species while others go extinct. Everything alive today has come forth from this synthesis of birth, change, and death. This is an audacious account of the origin of species. It opens up new imagination with regard to the natural world. All organic beings, living and dead, are related to one another, historically and biologically. All take their place in a single narrative of creative struggle, divergence, thriving, death, extinction, and further breakthrough. Common descent with modification by natural selection is the explanatory principle which interprets how species originate from one another, naturally. Our era reads this as a profoundly ecological insight.

A CROWD OF DIFFICULTIES

After presenting the main lines of the argument for evolution in its early chapters, *On the Origin of Species* sweeps forward through a wide range of biological topics. Much of the theory's value, Darwin argues, lay in the way it explains and unites so many different aspects of the natural world, from the fossil record to embryo development, from comparative anatomy to distribution of species around the world. First, though, the book stops its onward march and does a shrewd and startling thing. It raises serious objections to its whole argument:

> Long before having arrived at this part of my work, a crowd of difficulties will have occurred to the reader. Some of them are so grave that to this day I can never reflect on them without being staggered; but, to the best of my judgment, the greater number are only apparent, and those that are real are not, I think, fatal to my theory. (108)

Multiple problems are dealt with in chapters on broad difficulties, instinct, and hybridism. This is an astute move. Not only does it answer criticisms before they get lodged too deeply, but the answers in turn serve to show the explanatory reach of the theory across wide areas of biology. For our purposes we will consider briefly a few of the objections, lingering over how evolution could produce such a perfect complicated structure as the eye, and such a perfect instinct as honeybees display in constructing their mathematically precise honeycomb.

One problem that has surely occurred to the reader, observes Darwin, is the absence or rarity of transitional organisms. If natural selection works by choosing innumerable fine gradations, where are these intermediate forms? Why do we see well-defined species, and not find myriads of creatures in various stages of diverging from one species to the next? The main answer, in a word, is that in the slow process of evolution, the intermediate forms have gone extinct. Recall the diagram of taxa. Organisms better adapted in the struggle for existence gradually take over a habitat, and the parent form and transitional varieties will be wiped out by the very formation of the new form. That may be so, the reader may grant, but why do we not find an abundance of these transitional forms embedded as fossils in the crust of the earth? The answer lies in the imperfection of the geological record. In order for organisms to be preserved in layers of sediment, very precise conditions must be met, which situation happens only rarely. It is as if a collector went shopping for artifacts only periodically: "The crust of the earth is a vast museum; but the natural collections have been made only at intervals of time immensely remote" (173). This issue will be discussed later in depth.

Another sort of difficulty lies in how hard it is to envision transitions of animals with peculiar structures or habits. Take the case of the gradual conversion of a carnivorous land animal into one with aquatic habits, such as the seal. Opponents ask, "how could the animal in its transitional state have subsisted?" (179). Darwin admits that he lies under a heavy disadvantage with this issue, but adopts a strategy of looking laterally at other species to see how evolution might be imagined to have worked. First, *Origin* cites examples of species that bridge two ways of living. The weasel of North America, for instance, inhabits a semi-aquatic environment; during the summer it dives for fish; during the winter it leaves the frozen waters and preys on mice and other land animals. Its adaptation fits it for both habitats, which makes the long-term evolution of the seal more comprehensible. Or take the case of a land animal adapting to the air, such as bats which have gone from being terrestrial mammals to flying ones. How can this happen? Again looking laterally, squirrels can serve as models for such a transition. These animals have tails that range from slightly flat and skinny to wide and very close to the skin on their flanks. Let the climate and vegetation change or let stronger rodents immigrate into their territory, and the squirrels there would go extinct unless they improved in structure in some way. Individuals with fuller and fuller flank membranes could be modified and propagated until by the accumulated effects of natural selection the base of their tail unites with a broad expanse of flank skin, forming a parachute that enables them to glide through the air for an astonishing distance from tree to tree. The so-called flying squirrel is born, offering an analogy to the evolution of the bat.

To envision transitions, it is also helpful to think of a single organ that can be tweaked to great effect. The wings of birds, for example, are structured to give them the power of flight. But there are birds like the loggerhead duck who don't fly at all but use their wings solely as flappers; and like the penguin whose wings act as fins in the water and front legs on land; and like the ostrich whose wings function as sails; and like the water ouzel or dipper who uses its wings as underwater rudders. A structure can be repurposed, perhaps not perfectly, but just enough to allow its owner to be a bit better adapted to a new environment. Or take the case of flying fish, whose swimming pattern has them leave the water and glide through the air with the help of fluttering fins. If these fish had been modified in the struggle for existence into a winged animal, who would have ever imagined that in an earlier stage they had been inhabitants of the open ocean, and had used their incipient organs of flight exclusively to escape being devoured by other fish? Seeing as there are flying mammals, flightless birds, and animals that survive on both land and water, land and air, water and air, or all three together, it is not impossible to envision how transitions might take place over eons of time.

Yet another difficulty that might occur to the reader is the occurrence in animals and plants of organs of little apparent importance. Why would natural selection choose these? It would be well to keep in mind here, Darwin cautions, how ignorant we are in regard to what every organism needs to function well. The tail of the giraffe, a most trifling object at first glance, is a case in point, for it functions to swat flies, thereby allowing the giraffe to range into wider territory. But the question still persists in myriad other cases. The answer is that in the long process of evolution, organs of little apparent purpose were originally of high importance to progenitors.

When the situation changed, these organs would still be inherited though their usefulness would peter out. Clearly, if the organ became injurious to its owner, it would be modified, or else the species would become extinct, "as myriads have become extinct" (201). But if it just became useless, it would continue to exist, perhaps in a shriveled form, simply due to inheritance. Herein lies an argument against special creation, for why would a divine architect deliberately design such purposeless organs? "Hence every detail of structure in every living creature ... may be viewed, either as having been of special use to some ancestral form, or as being now of special use to descendants of this form – either directly or indirectly through the complex laws of growth" (200). This utilitarian idea flies in the face of natural theology which took it as a given that all things were created for a purpose relating to humans, some for utility, others to offer moral lessons, still others to please the senses: "They believe that very many structures have been created for beauty in the eyes of man." If such were true, it "would be absolutely fatal to my theory" (199). To the contrary, descent with modification makes clear that what we consider beautiful, such as the shape of an orchid, first and foremost has adaptive value to the species possessing the characteristic, not to human aesthetics. The structure of every living being, including vestigial organs, encodes an immense history.

Perhaps the most telling difficulty Darwin raises against his theory, one that continues to resonate in contemporary debates, is the existence of organs of extreme perfection. His chosen example is the eye. This is indeed a baffling problem:

> To suppose that the eye, with all its inimitable contrivances for adjusting the focus to different distances, for admitting different amounts of light, and for the correction of spherical and chromatic aberration, could have been formed by natural selection, seems, I freely confess, absurd in the highest possible degree. (186)

The dilemma arises because the eye is a composite of many individual adaptations, all of which must be present for proper functioning. This exquisite design makes an especially compelling case for special creation, since it is highly unlikely that each part could have evolved at the same time. People compare the eye to the telescope, a precise instrument designed by the direct action of the highest human intellects, and argue that the finished eye had to be created by a highly intelligent Designer in a similarly straightforward way. In a direct challenge to the theory of special creation, *Origin* queries, "But may not this inference be presumptuous? Have we any right to assume that the Creator works by intellectual powers like those of man?" (188).

Let us not be hasty in concluding that such a complex organ could not be formed by small, progressive, transitional gradations. Instead, look across the whole animal kingdom where a range of photosensitive organs makes it possible to envision how an eye could evolve through many intermediate steps. Imagine the eye, Darwin writes, beginning simply as a nerve that becomes sensitive to light. Envision this simple optic nerve covered over with a layer of transparent tissue. Then suppose every part of this layer to be continually changing in density, so as to separate into layers of different thickness, placed at different distances from each other and slowly changing form. Meanwhile, realize that there is a power intently watching each slight alteration, and carefully selecting any which in any way or any degree tend to produce a sharper image beneficial to its owner.

To digress: the eye could conceivably have evolved in stages, each stage making a minor modification to the previous one. A light-sensitive nerve would be beneficial, giving its owner the advantage of telling day from night in some way, or of sensing when a predator overshadows.

These cells multiply, and eventually selection would favor an arrangement that curves them into a concave surface; this would give the organism a rudimentary ability to sense the direction from which the light is coming. Eventually these cells and their curvature increase to form a cup with a pinhole opening, giving a greater sense of direction toward the light source. A clear layer of tissue forms that entirely encloses the opening; a rudimentary lens develops which can refract light to focus some sort of an image. More elements such as an iris and cornea form and are selected, producing a full complex eye.

The important thing to remember is that an organism does not necessarily need a perfect eye to benefit from sensitivity to light. At every stage, even one that we may consider less than perfect, the organ's working confers some benefit to its owner. There is no need to assume, then, that the eye would have to become a complex structure before it became functional. Instead, through a series of fine gradations it can continue to adapt ever more powerfully until reaching the exquisite perfection of the complex vertebrate eye. At every stage, Darwin writes:

> We must suppose the new state of the instrument to be multiplied by the million; and each to be preserved till a better be produced, and then the old one to be destroyed. In living bodies, variation will cause the slight alterations, generation will multiply them almost infinitely, and natural selection will pick out with unerring skill each improvement. Let this process go on for millions on millions of years; and during each year on millions of individuals of many kinds; and may we not believe that a living optical instrument might thus be formed as superior to one of glass as the works of the Creator are to those of man? (189)

As David Reznick observes, Darwin's answer to the difficulty posed to his theory by organs of extreme complication is that "such complexity arises through a process that is like climbing a long, winding staircase one step at a time, rather than leaping a tall building in a single bound."[7]

It is not only the beautiful entangled bank that has evolved, but the eye of the vertebrate creature that can contemplate it.

Beyond the evolution of anatomical structures, another set of difficulties involves the evolution of instinct, or an innate impulse that does not stem from learning. The social insects such as ants, bees, wasps, and termites form communities whose optimal function requires cooperative work including a division of labor among their members. Different members of the community seem to be born knowing their role. How does this instinctive behavior come about, if not through direct designations from the Creator? As with the exquisite structure of the eye, the solution is the same: natural selection also modifies instincts, privileging whatever gives an advantage. In this case the advantage is to the group, which then benefits its members: "selection may be applied to the family, as well as to the individual" (237).

Take, for example, the perfect geometrical construction of the honeycomb. It is a truly wonderful thing, made of a lattice of cells of the proper shape to hold the most amount of honey with the least expenditure of precious wax in their construction. Each cell of the honeycomb is hexagonal, optimally proportioned to fit with its six surrounding cells, and to nest with a back layer to produce a double-layered structure that makes the most efficient use of space and material. We hear from mathematicians that this structure solves an abstruse problem: how to pack the maximum number of cells of equal size into a given area. Yet this marvel is constructed by little insects of little intelligence. How can they make all the necessary angles and planes? And how do they know when it is done correctly and finished? "All this beautiful work can be shown, I think, to follow from a very few simple instincts" (224). Imagine that in ages past progenitor bees acted in simpler, less perfect patterns, making spherical cells, spacing them just so in a layered arrangement, extending the length of the cell as needed, making the walls neither too thick or too thin. Natural selection would act on these simple instincts as they became more refined because they confer immediate advantage:

> it is known that bees are often hard pressed to get sufficient nectar; ... a prodigious quantity of fluid nectar must be collected and consumed by the bees in a hive for the secretion of the wax necessary for the construction of their combs ... Hence the saving of wax by largely saving honey must be a most important element of success in any family of bees. (233–4)

Incremental improvement in construction of their comb would yield incremental advantage to the bees that made it. Natural selection would choose the swarm which wasted the least honey in the secretion of wax. This swarm in turn would transmit its economical habit to new swarms, which in their turn would have the best chance at success in the struggle for life. By taking advantage of numerous, successive, slight modifications of simpler instincts, natural selection would continue to privilege the bees' advantageous behavior until that "stage of perfection in architecture" (235) is reached where the comb is constructed with perfect economy of wax. This most wonderful of all instincts, that of the honeybee, can be explained by a law of nature.

The same type of reasoning can be applied to all the instincts of the social insects. One particular instance that would seem to be "fatal to my whole theory" (236) is that of the existence of sterile castes among the social insects, such as female worker bees or the sterile driver ants of West Africa. The instinct to sterility is a problem because it cannot be passed on to the next generation. Neuters do not breed. How is the instinct acquired, and to whose benefit does it operate? Again, natural selection offers a satisfactory explanation. Sterility evolved because it is good for the family. Successive, slight modifications to a few individuals kept them from reproducing but fitted them for certain services that benefitted the community; their reproductive energy was channeled into these tasks. Over time within the same nest, sterile castes became strikingly differentiated from each other by dissimilar size and structures of jaws, teeth, and the like, which enabled them to carry out specialized work. This high degree of division of labor and its attendant efficiencies are most useful to the community. "We can see how useful their production may have been to a social community of insects, on the same principle that the division of labour is useful to civilised man" (241–2). By continuously selecting the fertile parents who produce such offspring, that is, sterile workers fitted for precise tasks, natural selection could form a species which should regularly produce such neuters, so that the whole group would thrive.

The discussion of the evolution of instincts ends with a sobering after-thought. Darwin has been sensitive all along to the high cost of suffering paid by sensate animals if his theory is correct. Here he points to the cuckoo chick which evicts its foster-brothers from the nest, leaving them to die while it grows fat. He has studied ants which carry off other ant species to do their nest-cleaning and juvenile-rearing work, making them in effect slaves. Some species of wasps lay their eggs in the bodies of caterpillars; upon hatching, these feed on the bodies of the live caterpillars, which appear to be in distress, before chewing their way out into the world. On the theory of special creation, all this is according to divine design. "I cannot persuade myself," he wrote to Asa Gray, "that a beneficent and omnipotent God would have designedly created the Ichneumonidae [wasps] with the express intention of their feeding within the living bodies of Caterpillars, or that a cat should play with mice." Darwin finds it so much more satisfactory to look upon these instincts "not as specially endowed or created instincts, but as small consequences of one general law, leading to the advancement of all organic beings, namely, multiply, vary, let the strongest live and the weakest die" (244). It is somewhat of a relief to think of this law of nature as a secondary cause, in other words, rather than attribute these instincts to the direct plan of a loving God.

Having gone out of his way to lay out what evidence would count against his hypothesis, indeed, even be fatal to it, Darwin adroitly instructs the reader by showing how natural selection, rightly understood, can resolve the problems. Gaps in transitional forms, the evolution of complex structures such as the eye, the evolution of instinctive behavior such as building a honeycomb, even the evolution of sterile castes in social insects: difficulties which could have brought down the theory serve instead to give it more credibility. In the process, the explanatory power of the theory is displayed in new dimensions. Evolution works slowly. Nature tinkers. Living creatures and all their parts change, from simple to complex forms, according to what gives them an advantage. The words that end the discussion of the eye apply to the whole crowd of difficulties raised to this point and ring with quiet assurance: "If it could be demonstrated that any complex organ existed, which could not possibly have been formed by numerous, successive, slight modifications, my theory would absolutely break down. But I can find out no such case" (189).

With the theory well in hand and major difficulties answered, Darwin next takes his view of life deep into time and wide across space. In two chapters of extraordinarily powerful argumentation (9 and 10) *Origin* shows how descent with modification is explained and is supported by the fossil record, and in two more (11 and 12) demonstrates how the theory likewise makes sense of the present-day distribution of plants and animals across the globe. "We continually forget how large the world is" (302), an aside reminds us. The power of the theory to interpret life's long history and widespread geography paints evolution in true planetary colors.

THROUGHOUT TIME

By the mid-nineteenth century scientific knowledge of the earth and its deposits was booming. Geologists recognized that the upper crust of the earth was composed of layers of rock laid down in chronological sequence. Though not necessarily a series of neat stripes, because often jumbled by earthquakes and the like, these layers could be studied in order from the oldest at the bottom to the progressively younger near the top. (To use a familiar example, consider the Grand Canyon in Arizona. Its exposed walls reveal a sequence of sedimentary rock layers of different colors—red, ochre, tan—ranging downward in time to the dark Vishnu Schist at the bottom deposited approximately 1.7 billion years ago.) Paleontologists who studied the fossils found in the rock noted how species appeared, expanded, and disappeared within the different layers. They did not know the actual age of the different strata, just their relative position as older or younger. Once the connection was made between the sequential age of rock layers and the fossils they held, however, scientists started constructing a history of life, arranging species in order of chronological appearance, from fish, to amphibians, to reptiles, to mammals, from oldest to newest up through the rock. Drawing on this knowledge, Darwin puts his theory to work to interpret the past, reading both the fossil evidence and its many gaps. In the process, the theory delivers a staggering encounter with death and extinction, and powerfully confirms the relationship between the living and the dead in every region.

The grand vision of universal descent requires a grand vision of time. Darwin did not know the age of the earth in absolute years (current consensus dates the planet at 4.6 billion years old), but he starts the discussion of fossils with an appeal to geologic processes to show that earth is very, very old. This will ensure that when he places the story of life in this framework, there will be enough time for natural selection to work.

In the mid-nineteenth century there were two competing theories about how features of earth's landscape were formed. So-called catastrophism held that landscapes were shaped quickly due to violent calamities, such as the flood in the days of Noah or a huge wave gouging out a valley. The fast rate of change led to the conclusion that the earth was relatively young. The fossil record bore this out, showing that groups of animals and plants suddenly disappeared and were replaced by others. For those who held to special creation, this did not present a problem. The opposing view, uniformitarianism, championed by geologist Charles Lyell in *Principles of Geology*, argued instead that the earth's features were formed gradually by forces still active today. Over time the earth imperceptibly rises and falls, the same area being at one time under water and then lifted up to dry, even becoming a mountain with embedded seashells. Rain, wind, erosion, earthquakes, and the like then sculpt the surface features. Given the slow rate of change needed to build mountain ranges, excavate valleys, and carve seaside cliffs, it took millions of years to lay down rock layers.

Darwin loved this book. He had read it on the *Beagle*, and looked for and found evidence for its position during his travels. Now he uses its insights to argue that the earth is ancient beyond measure:

> It is hardly possible for me even to recall to the reader, who may not be a practical geologist, the facts leading the mind feebly to comprehend the lapse of time. He who can read Sir Charles Lyell's grand work on the *Principles of Geology*, which the future historian will recognise as having produced a revolution in natural science, yet does not admit how incomprehensibly vast have been the past periods of time, may at once close this volume. Not that it suffices to study the *Principles of Geology*, or to read special treatises by different observers on separate formations, and to mark how each author attempts to give an inadequate idea of the duration of each formation or even each stratum. A man must for years examine for himself great piles of superimposed strata, and watch the sea at work grinding down old rocks and making fresh sediment, before he can hope to comprehend anything of the lapse of time, the monuments of which we see around us. (282)

Darwin invites the reader to wander as he did along the seashore and observe the action of coastal waves eroding a cliff. The water reaches the cliff only twice a day at high tide; rock formations are worn down only "atom by atom." How unimaginably long this takes! Each pebble and subsequent grain of sand bears the stamp of deep time. Using published reports about the deposit of silt in the Mississippi delta, the thickness of the formations of the Weald in south-east England, and other data, he estimates it took hundreds of millions of years for features we see today to have formed. The lapse of time is astonishing. Trying to come to terms with it "impresses my mind almost in the same manner as does the vain endeavour to grapple with the idea of eternity" (285).

All along this passage of time, organisms emerge, mutate, diverge, and go extinct in a rhythm commensurate with the changing landscape.

> During each of these years, over the whole world, the land and the water has been peopled by hosts of living forms. What an infinite number of generations, which the mind cannot grasp, must have succeeded each other in the long roll of years! (287)

Now a difficulty, previously raised, receives fuller attention. By the theory of natural selection all living species are connected with the parent species of each genus, which, though now generally extinct, in their turn were similarly connected with more ancient species, and so on backwards, converging to the common ancestor of each great class. On this view, the number of transitional links between living and extinct species must have been inconceivably great. If there is any truth to the theory at all, assuredly such have lived upon this earth. Why then is there no trace of them in the fossil record? Why does the geologic record not yield a satisfying sequence of fossils that reveals the complete story of life over the endless roll of years?

The answer becomes clear if we understand how fossils are made. Fossils form when an organism dies and is quickly encased in sediment. The sediment forms a mold around the carcass, after which its bodily parts slowly dissolve away and are replaced by minerals. The fossil stays in the sediment as a mineral cast of the original creature. In this process soft tissues are seldom preserved; hard elements such as shells, bones, or teeth, or stems and woody material, lend themselves to mineral transformation. Because of the special conditions required, namely an ample supply of sediment or oxygen-free mud plus protection from disturbance over eons of time, most organisms that live never become fossils. They leave behind no trace in the rock matrix that they ever existed.

Geologists had concluded that the ideal locations for preservation were shallow seas in tropical or subtropical regions, since such seas have abundant life and experience much sedimentation. The most likely time for fossils to form is during the slow sinking of a region when sediment washing in from streams could accumulate in bays and deltas in sufficient quantity to infiltrate organisms as they died. The least likely time was when the land was rising upward, leaving no place for sediment to amass and exposing fossils to a gauntlet of destructive processes. Darwin's own observations lend concreteness to these general principles. Scarcely any fact struck him more when exploring the western side of the South American continent than the lack of fossil beds. He knew that geologists believed that this coast is slowly being upraised to form the Andes mountains, and saw the reason. For hundreds of miles along the whole west coast as soon as deposits are brought up by the gradual rising of the land, they are battered away by the incessant grinding action of the ocean waves. Future paleontologists will find no evidence of life here, though millions of creatures thrived on land and sea. The conclusion:

We may, I think, safely conclude that sediment must be accumulated in extremely thick, solid, or extensive masses, in order to withstand the incessant action of the waves when first upraised and during subsequent oscillations of level ... sediment may be accumulated to any thickness and extent over a shallow bottom, if it continue slowly to subside. In this latter case, as long as the rate of subsidence and supply of sediment nearly balance each other, the sea will remain shallow and favourable for life, and thus a fossiliferous formation thick enough, when upraised, to resist any amount of degradation, may be formed. I am convinced that all our ancient formations, which are rich in fossils, have thus been formed during subsidence. (290–1)

And now for the irony. Periods of uplift are more conducive to the formation of new species, with the environment opening up new niches for varieties to branch out and colonize; but during such periods there will generally be a blank in the geologic record. During subsidence habitat is destroyed and populations migrate away, leading to extinctions and far fewer species being formed. Yet it is these very periods that are most conducive to the accumulation of great deposits rich in fossils. "Nature may almost be said to have guarded against the frequent discovery of her transitional or linking forms" (292).

The explanation of how fossils are formed sheds light on why the record is necessarily intermittent. Some rock layers have rich fossil evidence; some have eroded remains; some have none at all, being laid down during an interval of non-sedimentation or uplift. Other factors that affect what we find in the geologic record include the migration of marine species in and out of an area; the fact that continents may have existed where oceans are now spread out, while present-day land masses may have once been under water (this, before today's knowledge of plate tectonics). There need be little wonder about the gap in the fossil record:

> If then, there be some degree of truth in these remarks, we have no right to expect to find in our geological formations, an infinite number of those fine transitional forms, which on my theory assuredly have connected all the past and present species of the same group into one long and branching chain of life. (301)

Although most assuredly during vast periods of time the world swarmed with living creatures, the imperfection of the geologic record means they are lost to our knowledge forever.

Those who think the geological record perfect will use it to reject the theory of descent by natural selection, Darwin knows. Against them he spins a magnificent literary metaphor taken from his friend and geological tutor: "For my part, following out Lyell's metaphor, I look at the natural geological record as a history of the world imperfectly kept, and written in a changing dialect" (301). Of this multi-volume history of life we have only the last volume; in this volume there are left only a few short chapters; in each chapter, only a few pages here and there; on each page, only a few lines; and on those lines, each word written in a different language which slowly changes throughout the interrupted succession of chapters. The fossil record is a broken-up book. The rocks do not tell the whole story.

Despite its incomplete condition, the geological record opens an irreplaceable panorama of life throughout time. Without forcing the evidence, *Origin* sets out to read the words that are available and to decode the blank spaces. According to natural selection, over the grand sweep of incomprehensibly long stretches of time, species slowly adapt to new conditions of life, are selected for advantageous changes, and reproduce offspring who inherit the new characteristics. These in turn parent progeny with yet further modifications, the whole line eventually diverging to form new species, while the original ancestor and less adapted cousins slowly go extinct. Darwin argues that this account generally matches the empirical patterns we find in the geological record.

To wit: New species appear in the rock strata very slowly, one after the other. Successive changes between them are many and gradual. Fossils in one region of the world show an affinity of structure with those above and below them in near layers of rock, more so than with fossils from remote layers. Not all species change at the same rate or in the same degree. Some do not become modified at all. These do not last, "for those which do not change will become extinct" (315). A species once lost never appears again in the geologic record, even if the same conditions of life should recur. The general rule seems to be a gradual increase in a species' number, till the group reaches its maximum; then gradual decrease and final disappearance, with rare exceptions. This accords more with natural selection than with some supernatural agency.

Matching the rocks to his theory, Darwin invites us to recall the diagram of taxa. Envision again how parent species A evolved over fourteen thousand generations to eight new species; remember how parent species B, C, D, and others became extinct. Lay this diagram up against the geological record, with the oldest layer of rock at the bottom. We find evidence of slow and scarcely sensible mutation leading to new species, and the extinction of myriads of older ones. The entire sequence in both of these media, the diagram and the geologic record, impresses itself on the mind "like the branching of a great tree from a single stem" (317).

To digress: like a strobe light, the intermittent fossil record enables us to read the words of life which are available from past eras. Scientific consensus today dates the oldest fossils of primitive life forms to 3.5 billion years ago; earliest fossil evidence of cellular life with a bound nucleus dates to approximately 1.8 billion years ago; multicellular fossils such as sponges appear in the record about 575 million years ago; the appearance of large, complex life-forms such as trilobites begins around 520 million years ago. In a series of occasional scenes from a slowly changing drama, the fossil record allows these words to declare from the depths of time: we were here, during this period, along with other organic beings in our community; we descended from those older ancestors, and gave birth to these newly modified forms; our descendants are connected to us in ramifying lines of generation.

And the missing volumes, absent chapters, blank pages, and dropped lines in the book of life? Modern researchers estimate that a complete inventory of all the species that have ever lived would number in the billions. Not only are these species never found in the fossil record, but almost all that are there have gone extinct. How to read the death and extinction that are so much a part of the story of life? The notion that all inhabitants of the earth have periodically been swept away by catastrophes does not hold much appeal. Instead, the study of rock formations reveals that "species and groups of species gradually disappear, one after the other, first from one spot, then from another, and finally from the world" (317). It is not that a species, like an individual, has a definite life-span, an internal clock that runs out. No fixed law determines how long any species lasts. Rather, facing unfavorable conditions or injurious agencies, none pre-planned but all coming about by complex contingencies, a species that does not adapt is reduced to fewer numbers. Rarity then leads to the probability of extinction, just as sickness precedes death in an individual.

The theory of natural selection sheds light on the dynamic process at play. Grounded on the belief that each new variety, and ultimately each new species, is produced by having some advantage in competition or cooperative ability with others in the struggle for life, it holds that the extinction of less-favored forms almost always follows. The better adapted species use up the prime resources to live and reproduce, leaving less vigorous forms bereft of the necessities for survival. Eventually they vanish. Domestic breeding provides a good analogy. When a new, slightly improved variety of short-horn cattle was raised in England, at first it supplanted the less-improved varieties in the same neighborhood; eventually it was transported near and far, even taking the place of other breeds in other countries. So too with nature: the appearance of new forms and the disappearance of old forms are bound together. Extinction is integral to the process of evolution.

Time and again *Origin* emphasizes the finality of the death of a species. Species once lost do not reappear. "When a species has once disappeared from the face of the earth, we have reason to believe that the same identical form never reappears" (313). "A group does not reappear after it has once disappeared" (316). "When a group has once wholly disappeared, it does not reappear; for the link of generation has been broken" (344). Even if a species with similar characteristics should occupy the same environmental niche, it would not be the same species, having evolved from different ancestors. Each species is a unique, unrepeatable budding of the tree of life. Most that have existed are gone forever. Darwin professes himself to be astonished at this phenomenon: "No one I think can have marvelled more at the extinction of species, than I have done" (318). Yet the empirical record of extinction is consistent with natural selection.

There is one other fact that supports the theory of descent with modification, namely, what Darwin for decades had called "the law of the succession of types," or "this wonderful relationship in the same continent between the dead and the living" (339). Living species are the budding twigs at the top of the tree of life; fossils are the underlying branches. In any given region, fossils are quite different from species wandering above ground, yet there are similarities. Australia is noted for its kangaroos and wallabies; fossil mammals from Australian caves are closely allied to the living marsupials of that continent. In Latin America, the gigantic armour of a fossilized extinct ancestor correlates even to the untrained eye to the structure of today's armadillos, sloths, and anteaters. The same relationship can be observed between extinct and living birds of New Zealand, flying birds and fossils from the caves of Brazil, extinct and living land-shells of the Madeira islands, and extinct and living brackish-water shells of the Caspian Sea, among others. On the theory of descent with modification, the long-enduring succession of the same types in the same areas is explained: "For the inhabitants of each quarter of the world will obviously tend to leave in that quarter, during the next succeeding period of time, closely allied, though in some degree modified, descendants" (340).

In a similar manner, the fossil record in each region gives evidence of the close relationship of extinct forms not only to their living descendants but also to others in the same rock layer. The longer back in time we go, the closer species approach one another in structure and function. Fish and reptiles, for example, having diverged from a common ancestor, display distinct affinities in their older forms. "Thus, on the theory of descent with modification, the main facts with respect to the mutual affinities of the extinct forms of life to each other and to living forms, seem to me explained in a satisfactory manner. And they are wholly inexplicable on any other view" (333).

By tracking the history of life back into deep time, *Origin* demonstrates the strength of the argument that all forms of life, ancient, recent, and now living, unroll through the eons as one grand natural system, linked by generation. Coming forward through successive intervals, divergence creates a blooming of intensely beautiful, different forms while extinction erases their ancestors. All the great facts of geology and paleontology plainly reveal the theory of natural selection to be a better explanation of the history of life than the common view of the immutability of species, associated with the special creation of each. If species are immutable, fossils would not necessarily show gradations of structure over time and in close proximity, as they do. If species arise via special creation, then the same species should be able to reappear again and again to occupy similar niches throughout history, which they do not. Natural selection makes more sense of the evidence.

While demonstrating the explanatory power of the theory, these geological chapters also alert the mind to marvels in the history of life: enormous intervals of time; irreplaceable extinctions; incalculable numbers of generations going forward; and profound affinities between the living and the dead.

ACROSS SPACE

The age of European exploration starting in the 15th century gave naturalists on that continent unprecedented opportunity to accumulate knowledge of the earth's flora and fauna. Collections such as the one Darwin made on the *Beagle* voyage were gathered in museums and universities to the point where a big picture of the global distribution of plants and animals became possible. By the time of the *Origin,* the study of biogeography was an emergent science. The new worlds were stocked with strikingly unfamiliar plants and animals. Whence the difference between European, Asian, African, American, and Australian organisms? Leading naturalists of the day interpreted the growing data as more evidence of design on the part of the Creator. As discoveries increased, they speculated on the number and locale of important "centres of creation" around the globe. *Origin* argues instead that the theory of evolution by natural selection gives a more plausible explanation of the geographic distribution of life-forms over the planet. The core theory works within the framing idea that species originate in one place from a common parent, then migrate and diverge, unless stopped by an impassable barrier.

Puzzling out global patterns of animal and plant distribution, Darwin begins by bringing three great facts to the reader's attention. First, environmental conditions do not account for the placement of species. Run a line north to south down the center of North America and down the center of Europe. Despite a common range of climate and geographic regions such as forests, marshes, mountains, and great rivers, the species found in similar habitats are different on both continents. Or circle the globe east to west in the southern hemisphere between latitudes 25° and 35°. In Australia, South Africa, and western South America we see a similar type of climate and land. However, the flora and fauna of one continent are utterly different from that of another: no koalas in Africa, no lions in South America, no armadillos in Australia. Clearly, a species' location is not due simply to a suitable environment.

The second great fact is that barriers of any kind have a significant impact on the distribution of species. The eastern and western shores of Central America are very close to each other, separated only by the narrow land bridge known as the isthmus of Panama; yet these Atlantic and Pacific coasts have hardly a fish, crab, or shelled animal in common. (Obviously, this was before the building of the Panama Canal). Similarly, on the eastern and western sides of the Andes mountain chain running like a spine down the far western side of South America the vegetation and animals are distinctly different.

A third great fact is the affinity of species with each other on the same continent or in the same sea. Travel from the Magellan plain at the southern tip of South America northward to the broad plains of La Plata. Along the route one can hear one birdsong yield to another with nearly similar notes, see nests constructed alike but with slight differences, find that one species of *Rhea*, a great ostrich-like bird, has been replaced by another closely related one. (Already in his *Beagle* journals Darwin had wondered why two of the most closely allied species should be found in the same country.) On these same plains as on the Andes mountains and in the surrounding waters, we find rabbits, hares, and various types of rodents. Unlike European animals of the same sort, however, they plainly display an American type of structure. The same holds true for nearby islands: "If we look to the islands off the American shore, however much they may differ in geological structure, the inhabitants, though they may be all peculiar species, are essentially American" (349).

These patterns cry out for explanation. On the one hand they are inexplicable if one supposes special creation: why duplicate efforts and create many distinct species to occupy otherwise identical habitats? Despite barriers, why not create the same species in waters and lands so near each other? Why congregate relatives in the same district? On the other hand, they are predicted by, make plausible sense of, or are at least consistent with the theory of descent with modification: "We see in these facts some deep organic bond, prevailing throughout space and time, over the same areas of land and water, and independent of their physical conditions. The naturalist must feel little curiosity, who is not led to inquire what this bond is" (350). According to the theory, the bond is created biologically in a process which allows each group of organisms to be traced back to a single species in a single place, which then migrates and diversifies.

> The bond, on my theory, is simply inheritance, that cause which alone,
> as far as we positively know, produces organisms quite like or, as we see
> in the case of varieties, nearly like each other. (350)

Over vast stretches of time accompanied by large climate and land changes, similar species which now inhabit the most distant quarters of the world originally proceeded from the same source population. Each species originated in one place, adapted to the soil, climate, and animal and plant life already there. Each then extended its range, adapting to new circumstances, sprouting new characteristics. Species come into being at one point on the earth, then migrate and evolve.

Once species migrate into an area, the slow process of descent with modification creates ever new and branching species. In their new homes they will be exposed to new conditions. They will experience pressure to adapt from the new physical environment and from biotic interactions with others in their mutual struggles for life, "the relation of organism with organism being, as I have often remarked, the most important of all relations" (350). Over time they will produce a succession of improved varieties and eventually species of even better-adapted descendants, or else go extinct. There is no law of necessary development in any of this, no pre-programmed outcome. From the range of developing possibilities natural selection will choose only those varieties which give benefit in each particular, changing situation.

This natural process explains the three great facts observed in the distribution of species. On the various continents the migrating organisms have evolved into different species; hence, despite conditions of life being nearly the same, the same landscapes in Europe and North America, or again across the southern continents, have different inhabitants. Barriers prevent migration, keeping a species that evolves on one side of a great obstruction in its place; hence the difference in Atlantic and Pacific mollusks separated by the narrow Central American isthmus. Affinity among species stocked within a region is due to their recent descent from a pre-existing common ancestor; the different South American Rheas are close kin. The explanatory power of the theory shows itself once again: "the simplicity of the view that each species was first produced within a single region captivates the mind. He who rejects it, rejects the *vera causa* of ordinary generation with subsequent migration, and calls in the agency of a miracle" (352). Recall that a "true cause" is recognized as having a real effective existence in nature, rather than being a hypothesis or figment of the mind.

Taking the bull by the horns in typical fashion, Darwin now brings up several problematic examples that challenge the lone center/migration pattern necessary for his theory to hold up. How can closely related species exist on the summits of distant mountain ranges? How can the same species exist in widely separated bodies of fresh water? How can the same or similar species exist on islands and on the closest mainland, though separated by hundreds of miles of open sea? These are extraordinarily important problems to resolve. Those who hold for multiple acts of special creation see no problem with the same species being created many times over, mountain top to mountain top, lake to lake, continent to island. According to the theory of evolution, however, it is impossible that identical species can be produced by natural selection from different sets of parents; the genealogy in each case would be different. Darwin has to show that the same species in different locations came from a single origin and subsequently dispersed. Then the theory would hold.

In broad strokes, *Origin* starts by laying out global factors that enhance or impede migration. Climate change is one, with an area that once served as a highway for migration becoming impassable over time. Change in the level of land and sea is another, with lower sea levels allowing for the emergence of land bridges over which animals and plants can pass from continent to island, or from island to island. What engages the author's attention much more intensely are what he calls "occasional" means of distribution, focused on plants. There must on occasion be long-distance dispersal across the ocean. Through extensive correspondence with other scientists and his own experiments in greenhouse, garden, fields, and local ponds, he tests whether and to what extent seeds can travel.

He took 87 kinds of small seeds, immersed them in jars of seawater for 28 days, then planted them; to his surprise, 64 germinated. Moving on to larger seeds, he floated them when they were green and juicy, and then when dry, finding that ripe hazel-nuts sank immediately but when dried they floated for 90 days and then germinated. An asparagus plant with ripe berries floated for 23 days, but when dried it floated for 85 days and then germinated. Combining such results with an atlas of the speed of

Atlantic currents, Darwin calculated that on average the seeds of 14 plants belonging to one country might be floated across 924 miles to another land before they sank or died. If they landed in a favorable spot, they would germinate.

Besides floating off on their own, seeds can also be transported by other means. They can catch on drifting timber or rafts of flotsam that carry bits of soil; they can survive in the crop of dead birds floating in the water (to his surprise nearly all seeds in the crop of a dead pigeon he floated in salt water for 30 days germinated); seeds can fly in the crop or gut of birds blown off course, and germinate after being excreted (he picked out 12 seeds from the excrement of little birds in his garden, and some germinated); seeds may get eaten by freshwater fish which are then devoured by birds and excreted far from home (he forced seeds into the stomachs of dead fish, fed these to eagles and storks, gathered their thrown-up pellets and excrements, and several seeds germinated); seeds ride in patches of earth attached to icebergs; they can get transported in dirt sticking to birds' feet. There is a limit to how far seeds can travel while retaining their vitality. This is why the flora of distant continents remain distinct. But given that all these means of transport have been in action year after year, for centuries and millennia, it makes sense that many plants are widespread.

Against this background, *Origin* addresses another difficulty. Darwin's ally, the botanist Asa Gray at Harvard, had established that kindred species of plants lived in snowy regions of the Alps and Pyrenees in Europe as well as, amazingly, at the summit of Labrador peaks in Canada and the White Mountains in the United States. It might make sense to conclude that the same species had been independently created at several distinct points. But the solution is simpler. During a recent geologic period Earth's temperature dropped; in the freezing cold weather, ice from the north polar region moved south in great sheets; temperate zones experienced an arctic climate. Darwin credits his contemporary, the Swiss scientist Louis Agassiz, with drawing vivid attention to this glacial age, whose remnants are still visible in gigantic moraines, erratic boulders, and scored and polished mountain flanks.

Consider that in the circumpolar regions (inside an arc encompassing today's Greenland, Canada, Alaska, Russia, Finland, Sweden, and Norway) the flora and fauna are remarkably similar around the world. As the cold came on and the glaciers slowly spread south, each southerly zone became fitted for the same arctic beings who moved south with the ice. These supplanted the temperate-zone creatures already there, who in turn moved further southward, unless they were stopped by a barrier, in which case they perished. Tropical plants retreated closer and closer to the equator and probably suffered much extinction. Former inhabitants of mountain peaks descended to the cold plains. By the time the cold had reached it maximum, the northern halves of Europe and the United States would be covered with arctic plants and animals. *Origin* invites us to imagine what happened next:

> As the warmth returned, the arctic forms would retreat northward, closely followed up in their retreat by the productions of the more temperate regions. And as the snow melted from the base of the mountains, the arctic forms would seize on the cleared and thawed ground, always ascending higher and higher, as the warmth increased, whist their brethren were pursuing their northern journey. Hence, when the warmth had fully returned, the same arctic species which had lately lived in a body together on the lowlands of the Old and New Worlds, would be left isolated on distant mountain-summits (having been exterminated on all lesser heights) and in the arctic regions of both hemispheres. (367)

With similarly far-ranging visualization Darwin charts the flow of species up and down the flanks of the Himalayas, on the mountains of New Zealand, along the Andes mountains in Chile, and in India, South Africa, Java, Japan, Australia, Borneo, and elsewhere, showing the migration of arctic, sub-arctic, and temperate forms in tune with changing climate. The intervals of time involved are exceedingly long. The land masses involved are enormously large, with specific plants and animals invading, mingling, and crossing borders throughout northern and southern, eastern and western hemispheres.

This narrative explains the existence of different but related species in alpine regions. Being surrounded by strangers, many migrants from the arctic will have to compete with new forms of life; advantageous adaptations will have profited them. Though still plainly related by inheritance to species in other regions, these wanderers now exist in their new homes as well-marked varieties or species. Acutely aware that this explanation does not remove all difficulties, Darwin is confident that it goes far enough to explain in a reasonable way the fact that the same alpine species can be found on widely separated mountain summits.

The story is different but the mobility of life still striking when the question shifts to the same species found in separated bodies of water. One might think that the intervening land would present an impenetrable barrier, but the opposite is the case. Allied species prevail in a remarkable manner throughout the world. "I well remember, when first collecting in the fresh waters of Brazil, feeling much surprise at the similarity of the fresh-water insects, shells, etc., and at the dissimilarity of the surrounding terrestrial beings, compared with those of Britain" (383). How can this be explained, if not by separate acts of creation?

The answer lies in organisms becoming fitted for short and frequent migrations from pond to pond, or from stream to stream, until a major barrier is encountered. Take fish as an example. The same species never occur in the fresh water bodies of distant continents or on opposite sides of a mountain range which separates river systems and thus prevents migration. But in contiguous areas, such is not the case. Floods, erosion, or slight changes in the elevation of the land can cause one stream to flow into another; stream capture opens up new habitats for a fish species to branch out. Or consider fresh water shell creatures, which have a very wide range. Experiments show that ducks or other birds might provide means of transport:

> I suspended a duck's feet, which might represent those of a bird sleeping in a natural pond, in an aquarium, where many ova of fresh-water shells were hatching; and I found that numbers of the extremely minute and just hatched shells crawled on the feet, and clung to them so firmly that when taken out of the water they could not be jarred off, though at a somewhat more advanced age they would voluntarily drop off. Those just hatched molluscs, though aquatic in their nature, survived on the duck's feet, in damp air, from twelve to twenty hours; and in this length of time a duck or heron might fly at least six or seven hundred miles, and would be sure to alight on a pool or rivulet ... (385)

With respect to water plants, wading birds which frequent the muddy edge of ponds are the most likely means of dispersal. Here Darwin recounts his experiment with seeds in the teacup of mud detailed in Chapter 1 above, adding that the same mechanism would also be in play for the dispersal of eggs of small fresh-water animals. When these get floated in a new body of water, the struggle for life may over time lead to new species, the new arrivals exterminating the native inhabitants who had not had to contend for resources up until then. After a time, we will find the same species in wide distribution over lakes, rivers, and streams. Pointing to J. J. Audubon who found the large seed of a water lily in the stomach of a great heron, Darwin figures that birds which have large powers of flight and naturally travel from one body of water to another are the main means of dispersal of fresh-water seeds and eggs. Not special creation, then, but "Nature, like a careful gardener, thus takes her seeds from a bed of a particular nature, and drops them in another equally well fitted for them" (388).

Coming to one last difficulty, the same or similar species found on continents as on islands far off shore, Darwin agrees that part of the difficulty lies in his theory's insistence that individuals of the same species have descended from the same parent, and therefore have proceeded from a common birthplace. The means of dispersal already discussed go part way to resolving the problem. Rather than dwelling on these again, *Origin* seizes the opportunity to press the explanatory power of his theory over the theory of independent creation in view of some other facts afford by island life.

The facts are these. The number of species on oceanic islands is scanty. Those that are found there are often endemic, exclusively native to that place and found nowhere else. Islands often possess odd trees belonging to orders which elsewhere include only herbaceous plants. No terrestrial mammals are found on islands further than 300 miles from the continent, but aerial mammals such as bats appear on almost every island. Whole orders are missing: batrachians (frogs, toads, newts) have not been found on any of the many islands with which the great oceans are studded.

Adherents of special creation can offer no reason for why a generous number of the best adapted plants and animals have not been created on oceanic islands, or for why the native creatures are so unusual. The herbaceous-order trees receive no rationale, nor are we told why the supposed creative force has produced bats that can fly but no land mammals on remote islands. As for frogs, "why, on the theory of creation, they should not have been created there, it would be very difficult to explain" (393).

All these facts, however, might be expected on the strength of the theory of natural selection. The low number of species is due to their need to migrate in from elsewhere. Their peculiar characteristics are due to their modification in one special place. In the absence of regular trees herbaceous plants might readily gain an advantage by growing taller and overtopping other plants, so natural selection, choosing for that advantage, would convert them first into bushes and then into trees. No terrestrial animal can be transported across a wide stretch of sea, but bats can fly across. Frogs are mostly absent because they are immediately killed by sea water and thus cannot survive migration.

Of all the characteristics of oceanic islands that Darwin's theory explains, the most striking is the affinity of island species to those of the nearest mainland. Here the grand example is the Galápagos Archipelago, situated below the equator about 600 miles off the west coast of South America. Almost every animal and plant on these islands bears the unmistakable stamp of the American continent. Follow the cadences of his reasoning:

> There are twenty-six land birds, and twenty-five of those are ranked by Mr Gould as distinct species, supposed to have been created here; yet the close affinity of most of these birds to American species in every character, in their habits, gestures, and tones of voice, was manifest. So it is with the other animals, and with nearly all the plants, as shown by Dr. Hooker in his admirable memoir on the Flora of this archipelago. The naturalist, looking at the inhabitants of these volcanic islands in the Pacific, distant several hundred miles from the continent, yet feels that he is standing on American land. Why should this be so? why should the species which are supposed to have been created in the Galápagos Archipelago, and nowhere else, bear so plain a stamp of affinity to those created in America? There is nothing in the conditions of life, in the geological nature of the islands, in their height or climate, or in the proportions in which the several classes are associated together, which resembles closely the conditions of the South American coast: in fact there is a considerable dissimilarity in all these respects. On the other hand, there is a considerable degree of resemblance in the volcanic nature of the soil, in climate, height, and size of the islands, between the Galápagos and Cape de Verde Archipelagos: but what an entire and absolute difference in their inhabitants! The inhabitants of the Cape de

Verde Islands are related to those of Africa, like those of the Galápagos to America. I believe this grand fact can receive no sort of explanation on the ordinary view of independent creation; whereas on the view here maintained, it is obvious that the Galápagos Islands would be likely to receive colonists, whether by occasional means of transport or by formerly continuous land, from America; and the Cape de Verde Islands from Africa; and that such colonists would be liable to modifications; – the principle of inheritance still betraying their original birthplace. (398–9)

Once landed, migrants to an island chain may evolve in dissimilar ways. Thus on the Galápagos, species on different islands vary from one another yet are still closely related. This is due not primarily to the physical conditions on the islands, soil, currents, and the like, but to the sets of other living inhabitants with which each species has to compete or cooperate in the struggle for life. A plant's seeds might find a spot on rich soil, or that prime spot may already be inhabited; they may be exposed to the attacks of somewhat different enemies; the circumstances bring about different modifications.

To digress: today the best-known example, which Darwin did not use, is the different species of finches whose beaks evolved on the various Galápagos islands to take advantage of the type of seeds available.[8] A multi-year study by Peter and Rosemary Grant documented how finch beaks differ from short, narrow, and shallow to long, wide, and deep, the differences in dimensions correlating with the birds' ability to harvest different types of seeds. In 1977 the islands experienced a severe drought; food was so scarce that no birds produced young that year. Only 15 per cent of the adults survived to reproduce when the rains finally came. Those that survived had longer, wider, deeper beaks that enabled them to crack the tougher seeds in the seed bank. Their offspring inherited the trait. In 1983 a continuing deluge of rain carpeted the islands with grass, providing an abundance of small, soft seeds. Finches with smaller beaks were more able to harvest the available seeds; many produced multiple sets of offspring. After this time the average bird in the population had a shorter, narrower beak, thus reversing the change that had occurred during the drought. As Reznick trenchantly observes, "This reversal is telling because it says that there is not a universal 'best' bird."[9] Whether a given feature of an individual gives it an advantage over another depends entirely on the circumstance, be it drought or flood. This is a good illustration of the fact that evolution does not progress in any particular direction, but is rather a response to present conditions at the moment. If conditions change, then so will the selection experienced by populations. If conditions remain constant, it is possible there will be no evolution at all.

The study of island life brought Darwin to the clear insight that inhabitants of oceanic islands migrate from a close land mass, then become subsequently modified and fitted to their new homes. This principle is of the widest application throughout nature. Regular patterns of distribution, unintelligible on the theory of special creation, now make sense. A mountain as it slowly rises is colonized from the surrounding lowlands. Lakes and marshes are related to creatures on the surrounding dry land. Blind species found in caves share characteristics with similar organisms above ground. These pervasive relations:

> are, I think, utterly inexplicable on the ordinary view of the independent creation of each species, but are explicable on the view of colonisation from the nearest and readiest source, together with the subsequent modification and better adaptation of the colonists to their new homes. (406)

Although admitting ignorance about the full effects of climate change, oscillating land and sea levels, glacier action, and means of transport, Darwin finds his theory full of light. The difficulties in believing that all individuals of the same species wherever located have descended from the same parents are not insuperable, even in the hardest cases of mountain summits, fresh water bodies, and oceanic islands. If we grant enormous periods of time for migration and modification, then the grand facts of the geographical distribution of species around the globe can be explained by this theory. We can see why two areas having nearly the same physical conditions should often be inhabited by very different forms of life. We can understand the importance of barriers which separate great geographical provinces of the world's animals and plants by preventing migration. We can understand the peculiar species of oceanic islands and their affinity to species on the nearest coast.

Origin ends its discussion of the distribution of species around the globe by connecting it back to the chapters on time and the geologic record. In both instances planet-wide evidence can be explained by the theory of evolution. Whether we look to the long succession of species read in the rock or to the continental spread of current life-forms, the theory brings a reasonable intelligibility to what we observe in the natural world. Throughout time and across space:

> in both cases the forms within each class have been connected by the same bond of ordinary generation; and the more nearly any two forms are related in blood, the nearer they will generally stand to each other in time and space; in both cases the laws of variation have been the same, and modifications have accumulated by the same power of natural selection. (410)

The importance of the geologic record and biogeography in providing crucial insights that support the reality of evolutionary change cannot be overestimated. Writing at his desk in Down House Darwin was still voyaging around the world, digging up fossils, checking out island species, forming a vision of living beings as one grand natural system linked by the bond of inheritance.

MUTUAL AFFINITIES

Following these historical and geographical surveys, the penultimate chapter of *Origin* probes the "mutual affinities of organic beings" by focusing on various branches of biology as then practiced. The overarching argument is that the problems encountered in these disciplines can be plausibly illuminated by the theory of descent with modification.

First up is the science of classification, which slots creatures into a spot on a chart composed of nested sets of groups ranked within groups from kingdom to species. By its very structure the classification table inherently recognizes that organisms are related in "chains of affinities." It would be preposterous, for example, to classify a kangaroo with a bear. Naturalists, however, ran into difficulty trying to figure out where to place animals and plants that didn't fit the usual mold. Unless Darwin is greatly deceiving himself, it is not external characteristics which most profoundly identify a species but its relationships of descent. Consider once again the diagram of taxa: "community of descent is the hidden bond which naturalists have been unconsciously seeking, and not some unknown plan of creation or the enunciation of general propositions, and the mere putting together and separating objects more or less alike" (420). All true classification is genealogical.

Comparative anatomy aims to study structural resemblances between species in order to understand similarities and differences (the similar configuration of bones in the hand of a man, the paw of a mole, the leg of a horse, the paddle of a porpoise, the wing of a bat). Rather than attributing such patterns to a plan of creation with its unchanging archetypes, the theory of common descent holds that such characteristics can be explained by inheritance from a common parent. In the above example, all the organisms mentioned are mammals which in their own ways carry forward the body plan of the mammalian ancestor. Comparing mammals with fish, however, would call up an immeasurably more distant past. While both have a backbone, the mammal did not evolve from the fish. Both fish and mammal start from a common point, a vertebrate ancestor, and each follows it own diverging road from there. Once again it is genealogy which sheds a clear light.

Comparative embryology studies the form of species prior to birth, noting similarities early in gestation and then differences of structure that emerge as the embryo matures. Darwin hewed to the common view that each individual embryo recapitulated all the stages of the history of its species as it developed, a view that is now discredited. The gist of his argument, however, holds firm, that descent with modification can explain similarities across species before birth. In this section Darwin the observer steps forth with an extraordinary range of references to the embryonic and adult forms (presented here in *Origin*'s sequence) of moths, flies, beetles, thrushes, cats, lions, the gorse plant, acacias, mammals, chicks, tadpoles, blackbirds, barnacles, butterflies, bats, porpoises, cuttle fish, insect larvae, cattle, human children, silk moths, greyhounds, bulldogs, cart horses, race horses, pigeons (again!), spiders, and aphids. Among related species the interesting empirical patterns observed in embryology reveal community of descent, the early embryo being "a picture, more or less obscured, of the common parent-form of each great class of animals" (450).

The study of rudimentary or atrophied organs, of no earthly use to their owner yet extremely common throughout nature, also finds a sensible interpretation in his theory. Embryonic ruminant calves have teeth that never break through the gums; embryonic flightless beetles have shriveled wings. Almost incomprehensible on the theory of special creation, these organs can be seen as part of an organism's inheritance. Though no longer useful, they are passed on as a fading echo of the life led by some distant ancestor. The eyes of animals in dark caverns, the wings of flightless birds, the stump of a tail, the vestige of an ear: rudimentary organs may be compared with the letters of a word which are still retained in the spelling even though useless in the pronunciation (think of the word "through"). While no longer useful, at least they serve as a clue in tracing the word's derivation. On the view of descent with modification, *Origin* observes that the existence of organs in a rudimentary, imperfect, or quite aborted condition, far from presenting a strange difficulty as they assuredly do on the ordinary doctrine of creation, might even have been anticipated, and can be accounted for by the laws of inheritance.

Major issues in these various biological disciplines fall into place on the view that descent with modification has left an imprint on all aspects of living organisms. This single explanatory framework gives rise to scientific explanations at once more precise and more encompassing. With it we can clearly see that all living and extinct forms are grouped together in one great natural system, with groups nested within groups, carrying the imprint of their history. In a remarkable final sentence, Darwin asserts that the evidence of this chapter is so compelling that it would convince him of common descent even apart from all his other reasons:

> the several classes of facts which have been considered in this chapter, seem to me to proclaim so plainly, that the innumerable species, genera, and families of organic beings, with which this world is peopled, have all descended, each within its own class or group, from common parents, and have all been modified in the course of descent, that I should without hesitation adopt this view, even if it were unsupported by other facts or arguments. (457–8)

"THERE IS GRANDEUR IN THIS VIEW OF LIFE"

With as many cards now laid on the table as he can afford in this "abstract," Darwin moves to a grand finale. Using another good pedagogical tool, the concluding chapter opens with the promise of a summary: "As this whole volume is one long argument, it may be convenient to the reader to have the leading facts and inferences briefly recapitulated" (459). In David Reznick's creative conceit, this summary unfolds as if Darwin were an attorney arguing in court for the acquittal of a client. "He probably envisioned himself facing a jury box filled with Cambridge dons, mentors, role models, and colleagues, including John Herschel, William Whewell, Adam Sedgwick, Richard Owen, Louis Agassiz, Karl von Baer, Asa Gray, Charles Lyell, Joseph Hooker, Thomas Huxley, Robert FitzRoy, and George Cuvier (in spirit), all leaning forward, staring at him intently, most shaking their heads in disapproval."[10]

In the opening statement lawyer Darwin sets out an overarching defense. Given the range of evidence natural selection can interpret, this law of nature is as central to the study of life as the law of gravity is to astronomy and physics. Reviewing next the case for the prosecution, he summons the most powerful arguments against his own position that he can muster: "Nothing at first can appear more difficult than to believe that the more complex organs and instincts should have been perfected not by means superior to, though analogous with human reason, but by the accumulation of innumerable slight variations, each good for the individual possessor" (459). Nothing can be more difficult to believe, in other words, than that the design we see in the natural world has not been produced by special acts of creation but is the result of the working of the natural world itself. Darwin argues that his work has endeavored to give these grave objections their full force.

Nevertheless—now he switches gears to the case for the defense— difficulties can be met if we admit the existence in nature of variations, the interactions of organisms with one another in competition or cooperation for the good of life, the preservation of each slight profitable deviation, the handing on of each advantage through inheritance, and abysses of time over which divergence and extinction slowly, slowly, slowly shape species in organic relationship with one another. He recaps the evidence from geology, geography, and the other disciplines, arguing that his theory's value lay in the way it explains and unites so many different features of the natural world. All these grand facts "are utterly inexplicable on the theory of independent acts of creation" (478) but make sure sense through the theory of common ancestry.

Moving to the summation, he states forcibly that observation and reason have convinced him that species slowly change, one giving birth to another: "all the organic beings which have ever lived on this earth have descended from some one primordial form, into which life was first breathed" (484). But not many agree. Why, he asks rhetorically, have all the most eminent living naturalists and geologists rejected this view of the mutability of species? One simple reason is that we are slow to admit any great change where we do not see the intermediate steps. Another, pertaining to our human limits, is our inability to grasp time as long as one hundred million years: our mind cannot add up and perceive the full effects of many slight variations, accumulated during an almost infinite number of generations. Yet another reason is political and psychological; it lies in the fact that senior naturalists, "whose minds are stocked with a multitude of facts all viewed, during a long course of years, from a point of view directly opposite to mine," (481) have too much invested in their own position to be open to new ideas. To this degree Darwin grants the position of his opponents a modicum of understanding.

Yet these same senior naturalists, who demand a full explanation of every difficulty that accompanies the theory of evolution, seem no more startled at a miraculous act of creation than at an ordinary birth. The lawyer's rhetoric flourishes as he presses opponents on specifics:

> Do they really believe that at innumerable periods in the earth's history certain elemental atoms have been commanded suddenly to flash into living tissues? Do they believe that at each supposed act of creation one individual or many were produced? Were all the infinitely numerous kinds of animals and plants created as eggs or seed, or as full grown? and in the case of mammals, were they created bearing the false marks of nourishment from the mother's womb (i.e. belly button)? Although naturalists very properly demand a full explanation of every difficulty from those who believe in the mutability of species, on their own side they ignore the whole subject of the first appearance of species in what they consider reverent silence. (483)

It is so easy to hide our ignorance under the idea of the "plan of creation" and to think we give an explanation when we only restate a fact. In a final argument against the advocates of special creation, Darwin tries to persuade:

"To my mind it accords better with what we know of the laws impressed on matter by the Creator, that the production and extinction of the past and present inhabitants of the world should have been due to secondary causes, like those determining the birth and death of the individual" (488). Lest any should take offense and think this demeans the natural world, his next words glow with affirmation: "When I view all beings not as special creations, but as the lineal descendants of some few beings which lived long before the first bed of the Silurian system was deposited, they seem to me to become ennobled"[11] (489).

Knowing that the jury will likely return a verdict of guilty and his theory will be rejected by most experienced naturalists, Darwin ends his defense with a determinedly hopeful look to the future. Once we see that organisms have a *history*, when we understand that all living beings are *related*, then the light will dawn. A new generation of young naturalists will rise to this theory, giving it a fair hearing. There will be a thrilling revolution in natural history; new methods of study and fields of inquiry will open up. Down the road, "Light will be thrown on the origin of man and his history" (488). Beyond these particulars, the theory itself is a bearer of hope. Regardless of its probable rejection for the moment, evolution will continue for unfathomable periods to bring forth creatures ever more perfectly adapted with bodily and mental endowments. The conclusion of this great peroration opens with the now famous image:

> It is interesting to contemplate an entangled bank, clothed with many plants of many kinds, with birds singing on the bushes, with various insects flitting about, and with worms crawling through the damp earth, and to reflect that these elaborately constructed forms, so different from each other, and dependent on each other in so complex a manner, have all been produced by laws acting around us. (489)

These laws, whose explication and defense form the substance of the book, include variation, inheritance, a rate of growth so high as to lead to a struggle for life, and natural selection, entailing divergence and extinction. Out of this process, with its enormous toll of destruction, great beauty comes; out of this natural process replete with death, the great good of the production of the higher animals directly follows. The case for the defense rests with a final sentence that rings down through subsequent years with eloquent appeal:

There is grandeur in this view of life, with its several powers, having been originally breathed into a few forms or into one; and that, whilst this planet has gone cycling on according to the fixed law of gravity, from so simple a beginning endless forms most beautiful and most wonderful have been, and are being, evolved. (490)

Thus ends this groundbreaking work, on a note of cosmic beauty and wonder.

Packed with factual information, studded with richly inventive metaphor, *Origin*'s one long argument comes into focus, in retrospect, with piercing clarity. Literary analysis has shown how Darwin's voice as a writer, by turns persuasive, friendly, dazzling, humble, dark, and warmly human, invites readers to use their own image-making powers to see the grandeur in this view of life that the author himself envisions. His writing has been compared to the novels of Victorian contemporaries Charles Dickens in *Great Expectations* and George Eliot in *Middlemarch*, authors who wove stories with complex, interlacing lines into a single overarching narrative. By involving the reader in the narrative experience at once tragic, awesome, and mundane, the book functions as literature. It is as a work of science, though, that *Origin* has had its most significant impact. Its "interlocking double-punch"[12] of lavish massing of scrupulously considered data coupled with compelling interpretation has set out one of the most impressive proposals in the history of science. It remains a groundbreaking treatise for the contemporary discipline of biology. Equally, it stands as a watershed for human awareness, profoundly altering our understanding of the natural world and, just as profoundly, of our own membership in the evolving community of life.

4

EVOLUTION OF THE THEORY

If the landscape reveals one certainty, it is that the extravagant gesture is the very stuff of creation. After the one extravagant gesture of creation in the first place, the universe has come to deal exclusively in extravagances, flinging intricacies and colossi down aeons of emptiness, heaping profusions on profligacies with ever-fresh vigor. The whole show has been on fire from the word go.

Annie Dillard

THE CENTER HOLDS

On the Origin of Species is a sustained argument showing that all living beings on Earth are related through common descent from simple ancestors, their diversity explainable as a result of natural processes. In none of *Origin*'s later revised editions did Darwin budge an inch from this original idea. Species originate by the action of natural laws, not supernatural acts of special creation. They change, rather than remain immutable. They branch out geographically from an original ancestor, rather than appear suddenly in different centers of creation. Because extinction breaks the bond of generation, species that disappear never return, rather than showing up in new acts of creation again and again. All living and dead organisms are profoundly related to each other in one grand natural system, rather than being separate creations related directly to the Creator but not to each other. That all of this has lasting value, he remained convinced.

The result of evolution thus envisioned is staggering. While the more than 1.5 million species of plants and animals already identified on this planet vary enormously in size, function, and their adaptation to all corners of the planet from blazing hot desert to icy tundra, from deep undersea to the top of Mount Everest, all are related members of the great tree of life. Mammals including humans are descended from shrew-like creatures that lived more than 150 million years ago. Mammals, birds, reptiles, and fishes descended from worm-like creatures that lived in the oceans 600 million years ago. These animals plus plants descended from bacteria-like micro-organisms that originated more than 3 billion years ago. The story that Darwin discerned tells how all these families of species gradually appeared on Earth over endless generations by a complex, unpredictable process kept on track by natural selection.

Stepping back from the immediate narrative, scientists have noted key features that characterize evolution as a whole. Kinship is a most striking one: all living beings on the planet are interrelated by common descent. The emergence of novelty is another impressive element: new forms of life never before seen appear in the course of time with new properties and abilities amid new networks of relationships. Cumulative bodily relationship is yet another: these new life-structures are not assembled from scratch but take shape through modifications made to already-existing simpler forms, to the point where the accumulation of many small changes leads to new organs with advanced form and functions (recall the eye), and eventually even to new species. Death is another feature of the story, a sobering companion of this biological creativity. In a finite universe the logic is inescapable: new patterns can only come into existence if old ones dissolve to make place for them; "new forms of life only through the death of the old."[1] Seen in retrospect, a trend toward complex organization also characterizes the process. While evolution wanders, diverges into dead-ends, indeed does not aim at any goal beyond successfully fitting an organism to its surroundings at any particular moment, its results over time show an in-built propensity to produce beings of ever more complicated structures by elaborating on simpler structures that already exist. Once life ignites from inorganic matter, living creatures evolve to the point of being conscious and then self-conscious, each capacity a function of increasingly organized nervous systems and brains. The dynamic of evolution, continuing still, works quietly over deep crevasses of time to bring forth the living world of nature in its ever more beautiful diversity.

Since its strong presentation in *Origin*, the theory of evolution has stood firm through many empirical testings. Today there is no reasonable scientific conflict about its reality. Details are still debated and investigated. But the premise that evolution renders a reliable account of how the living world came to be over 3.5 billion years of Earth's history forms the central organizing principle of the discipline of contemporary biology as practiced by scientists the world over. Supported by evidence from multiple sub-disciplines, the theory is held to be reliable in a way similar to the view that Earth and the other planets revolve around the sun. Borrowing a legal term, the theory of evolution, which explains how all living organisms developed by descent from common ancestors, is now judged to be accurate beyond a reasonable doubt.[2]

In the years following *Origin*'s appearance, however, ongoing scientific discoveries have continually nuanced understanding of precisely how evolution works. To dialogue with the theory as currently understood, we need to trace updates on several fronts. Toward that end, this chapter first flags the social misuse of the theory known as social Darwinism. It then concentrates on recent information that adds fine distinctions to the theory itself in a biological context. Finally, it pulls back to show how the interpretive frameworks of cosmology and ecology continue to give the theory legs.

MISUSE OF THE THEORY

Darwin's theory of evolution by natural selection is from beginning to end a scientific explanation crafted to account for the biological diversity of life as we see it today. Based on innumerable observations and ingenious abstractions, it interprets the diversity of species to be the result of natural causes rather than supernatural ones. Before long, however, this biological theory was transferred into the social realm and made to justify a slate of political, economic, and cultural agendas. It is essential to realize that such usage is widely repudiated by scientific practitioners today. The fatal flaw lies in transferring a scientific theory about a biological process that happens without conscious intent to the arena of human interaction where causes are intentional, willful, and complex.

The metaphor of "struggle for existence" in *Origin* expresses the interplay of energies among organic beings in an interdependent setting. *Origin* takes the trouble to explain the varying senses in which this plays out, whether by competition, cooperation, ingenious adaptation, or other means of exchange. As organic beings interact with each other and their physical environment, natural selection privileges whatever advantage happens to develop. These advantages occur randomly, to use contemporary language. However, once the fifth edition of *Origin* adopted the phrase "survival of the fittest," a concept coined by the English political theorist Herbert Spencer, the book's thesis was unfortunately linked with Spencer's philosophy of inevitable progress that entailed social winners and losers among human beings. By the turn of the twentieth century, a loose collection of ideologies had gathered under the label "Social Darwinism." These theories generally held that the powerful in society are innately better than the weak; that their success is proof of their superiority; and that social progress requires action toward specific goals that control the less fit, or even eliminate them. The eugenics movement which advocated selective breeding of human beings drew on these views. So too did practices of discrimination against certain racial groups such as in New Zealand, where Darwin's theory was invoked to justify the suppression of the indigenous Maori people. Nazi ideology that regarded non-Aryan peoples, the Jewish people in particular, as inferior was another vicious instance. A biological theory about the emergence of species was used to justify human atrocities.

Diane Paul's study of the social perversion of Darwin's theory notes how contradictory the results have been.[3] Whether conservative or liberal, people have found in evolution a language to promote views about human society which they already held. At first Karl Marx called the *Origin* "absolutely splendid," thinking it provided a basis for his theory of class struggle. After a time he was glad to get beyond it, however, seeing as it did not hold up the idea of progress toward a precise goal. At the opposite end of the economic spectrum, promoters of laissez-faire capitalism found the *Origin* supportive of their practice of unregulated competition for profit, where the wiliest would succeed. Militarily, some used survival of the fittest to argue for war, since the strongest nation would prevail. Others drew on the theory to promote peace, reasoning that because war kills young men in their prime, it prevents them from reproducing and thus weakens the nation. On and on it went. Colonial conquest vs. anti-imperialism, liberalism vs. socialism, patriarchy vs. feminism: Darwin's biological ideas became a resource for advocates of diverse political and social causes. What is wrong with this picture is that it makes a serious category mistake, taking a biological theory to be a script for human society.

Darwin's own record in this regard is ambiguous. From his youth he had imbibed the values of the Darwin and Wedgwood families which were famously abolitionist and promoted social reforms to aid the poor. Encountering the slave trade at first-hand during the *Beagle* voyage, he wrote critically and with feeling about the European treatment of indigenous peoples. A brief passage in *Voyage of the Beagle* is indicative: while traveling in Brazil he rode past a massive, steep hill of bare granite. His guide pointed out that in the past, runaway slaves had managed to eke out an existence by cultivating a little ground near the top. Darwin recounts:

> At length they were discovered, and a party of soldiers being sent, the whole were seized, with the exception of one old woman, who, sooner than again being led into slavery, dashed herself to pieces from the summit of the mountain. In a Roman matron this would have been called the noble love of freedom; in a poor negress it is mere brutal obstinacy.[4]

Note the sarcastic irony. Darwin saw no rational basis for subjugating other races to the Caucasian race because all humans belong to the same species, being descended from a common ancestor.

Yet in later writings he did indicate that differences in emotional and intellectual powers could be inherited, and indeed were inherited by persons according to their gender, race, and class. Educated white males of the upper class in imperialist Anglo-Saxon nations held a privileged place in the ongoing competitive march of civilization, while women, black people such as Australian aborigines, and poor people were clearly unequal in their capacities. While acknowledging the legacy of concern for the poor, weak, and unfit in the Jewish and Christian traditions, he saw the need for those more gifted to succeed in society. Thus he did not explicitly oppose the marriage others began to make between evolutionary theory and social policy during his lifetime, and tended in some of his later writings to contribute to this new way of justifying prejudice. In the struggle for resources, he thought it was important that the strong survive for the good of the human species as a whole.[5]

Regardless of Darwin's own ambiguous position, opposing slavery while holding to the superiority of the white man (*sic*) being one egregious example, today's scientific interpreters condemn social use of his theory as an ignorant and odious misappropriation. That the error continues to need correction can be seen from the exhibit mounted by the American Museum of Natural History in New York in 2009 to celebrate the 150th anniversary of *Origin*'s publication. One panel is entitled "Misusing Darwin's Theory" and reads in part:

> Darwin passionately opposed social injustice and oppression. He would have been dismayed to see the events of generations to come: his name attached to opposing ideologies from Marxism to unbridled capitalism, and to policies from ethnic cleansing to forced sterilization. Whether used to rationalize social inequality, racism or eugenics, so-called Social Darwinist theories are a gross misreading of the ideas first described in the *Origin of Species* and applied in modern biology.[6]

Darwin construed his work in *Origin* as an investigation of the natural world, not a political tract. His ideas are descriptive of the evolution of life, not normative for intentional human behavior. To assume otherwise is to transfer an argument from one discipline to another where it has no experimental basis. Such a category mistake ends up in sloppy thinking, to say nothing of disastrous social policies.

SCIENTIFIC ADVANCES

In the early part of the twentieth century the theory of evolution, which up until then had been making its way with mixed success in scientific circles, received a major boost when the laws of heredity were wedded to Darwin's own work. Gregor Mendel, a contemporary of Darwin, was an Augustinian monk who lived in what is now the Czech Republic. Through years of work in the monastery garden he discovered that the inheritance of certain traits like the color and height of pea plants followed particular patterns, depending on the combination of dominant or recessive genes. Most naturalists at the time, including Darwin, thought offspring received traits that were blended from both parents. Mendel figured out that while blending occasionally occurs, the standard pattern is that hereditary traits do not combine but are passed on intact; each parent transmits only half of the hereditary factors received by an offspring; different offspring of the same parents receive diverse combinations of these halves, which create different wholes.

Mendel's laws of heredity are the foundation of the science of genetics. Joining his understanding with the idea of natural selection gave the theory of evolution a new lease on life. In the *Origin* Darwin had lamented his ignorance of the laws of reproduction which left a gap in how descent with modification actually works. Now an account was to hand: genes in diverse combinations bring forth the heritable variations which are either favored or discarded by natural selection. Referred to as the modern synthesis or neo-Darwinism, this understanding was subsequently deepened by ongoing refinements such as the model of the double helix structure of DNA, the molecular basis of the genome. Mutations at the molecular level cause changes in the genetic information passed on to offspring. At this granular level, "the fundamental evolutionary event is a change in the frequency of genes and chromosome configurations in a population."[7] Such alterations increase or decrease the percentage of carriers of different genes, with corresponding changes in anatomy or behavior that fit the species for better or worse adaptation. By the 100th anniversary of *On the Origin of Species* in 1959, the modern synthesis (evolution by natural selection + genetic theory + its molecular detail) was firmly established.

Since then, further discoveries have strengthened the credibility of what Darwin proposed. Paleontologists have uncovered hundreds of fossils which show transitions from ancient to modern taxa: from fish to amphibians, reptiles to mammals, dinosaurs to birds, and terrestrial mammals to whales, though, as Darwin emphasized, the record will never be complete. Contemporary radiometric dating of earth's layers has been correlated with fossils found therein to arrive at species' ages, not just relative ages as in older or younger, but their actual ages in real time.

This information is correlated with family trees constructed from DNA sequences to estimate the time periods when lineages diverged from one another. New fields of research such as population genetics and the use of statistics to analyze variation bring forth yet more evidence that explains how the process works. The modern synthesis and its investigative branches have updated the core theory of evolution, keeping its explanatory power relevant and fresh.

At the same time, a number of scientific positions now differ from certain secondary claims made in *Origin*.[8] The discontinuity in details adds important nuance to contemporary understanding.

Speed

One major change concerns the length of time required for evolution to occur. Given that the adherents of special creation posited the instantaneous appearance of species, Darwin emphasized over and over again how slow the process was. "Nature does not make leaps," he wrote numerous times, meaning changes happen only gradually. While that is ordinarily the case, biologists now understand that the process can take place much more quickly. An intriguing example comes from the species known as the peppered moth in England. In the 1840s the vast majority of these moths had a light, patchy coloration; a few were dark-colored, but were so rare as to be a collector's item. By the 1920s, however, the dark moth was nearly the only kind that could be found in and around some cities. The cause? Natural selection working amid growing industrialization. The smoky exhaust of factories stripped trees of their mottled lichen and turned trunks and branches black. The light-colored moths, long camouflaged by the blotched bark, became easy prey for the birds when seen against the black. With the advantage of their color, however, the black moths flourished as the trees darkened from pollution. In the end, they became predominate. This change took less than eighty years.

David Reznick offers another fascinating example from his own research with guppies on the Caribbean island of Trinidad. Working in streams that consisted of pools separated by little barrier waterfalls, he tested the theory that in sections of streams with predators and hence with a higher mortality rate, natural selection would favor guppies that mature early and produce more offspring; delayed maturity and fewer offspring would mark guppies living a more relaxed life in pools with fewer predators. With this assumption, Resnick took guppies from a high predation pool and placed them in a pool with no predators at all. At the start, all the guppies were the same, genetically speaking. Over time, the little fish in the predator-free environment grew larger, reached maturity later, and produced fewer offspring. The reverse process took place when predators were introduced to a previously safe pool: reproduction took place at a younger age and produced more offspring. Combined field and lab work showed that new species had evolved in both places. Reznick speaks to the time involved:

> When I described this experiment to other biologists before I did it, I often saw sympathetic smiles and was told they hoped I would live long enough to see something happen. It turns out that I did not have to wait long. Males changed completely within four years, which is six to eight generations. Females began to show significant change in seven years.[9]

The fish in his experiments had evolved at a rate that was 10,000 faster than what was considered to be rapid evolution in the fossil record. There are multitudes of such examples. Adaptation and speciation can occur at a much faster rate than Darwin had assumed. Evolution may continue to happen rapidly right under our noses.

Catastrophe

By the late twentieth century scientific data had convinced geologists and paleontologists that the community of life likely suffered mass extinctions at five different points in the past. Thus was reintroduced a new form of the old catastrophe theory, though this time with a natural rather than a supernatural explanation. Dated roughly 450, 350, 250, 200, and 65 million years ago, these massive die-offs were due to continental break-ups, dramatic climate shifts, or the impact of asteroid hits. Today's thinking is hospitable to the idea, for example, that 65 million years ago an asteroid hit Earth, bringing about the extinction of the long-lived dinosaurs along with more than half the animal and plant species exiting on the planet. During these events multiple species disappeared not gradually but all at once, and not only locally but over far-ranging areas. Species went extinct not because they lacked an advantage over others in the struggle for life but because they could not cope with brutal disturbances in their physical environment. The end result was widespread and rapid decrease in the amount and diversity of life on Earth, which then took millions of years to reconstitute itself. Thus while Darwin was right about the gradual extinction of species in general, the reasons for extinction cannot be subsumed into a single causal theory, as *Origin* assumed.

Rhythm

The fossil record presented Darwin with evidence that evolution happens continuously by a series of slow, steady transformations. Proposed in the 1970s, the theory of punctuated equilibrium holds to the contrary, that evolution tends to be characterized by long periods of stability, or equilibrium, punctuated by episodes of very fast development. Most species will exhibit little change for most of their geological history. When change does occur, it will happen explosively in multiple branching events over a comparatively brief period of time. Then evolution will go somnolent again for another long era. This theory about the rhythm of evolution is still being disputed.

Struggle

Almost all commentators follow Darwin himself in crediting his reading of Malthus for triggering the idea of the struggle for existence. Over-abundant fecundity leads to many individuals competing for resources; in this context, increased variations bear great potential for change. Today's science has demonstrated that natural selection can function even in a population with no increased numbers to check. Many factors other than struggle are involved, such as a species' better integration with the environment, more efficient utilization of available food, better care of the young, and more cooperative social organization. So long as there is variation, evolution will occur.

Plate tectonics

In today's thinking, a great continent called Gondwanaland once existed, comprised of what are now land masses in the southern hemisphere: Antarctica, South America, Africa, Madagascar, India, Australia, and New Zealand. Starting about 200 million years ago this super-continent began to split apart and its pieces were carried away on different continental plates. Similar fossils found in similar geologic rock formations in these different parts of the world can be explained by the mechanism of plate tectonics. Migration of species took place in a much wilder way than Darwin's explanations for the geographic distribution of species ever imagined.

Different life forms

The tree of life on which lineages split but never rejoin may not extend all the way down to the base. Certain bacteria, rather than reproduce in the accustomed manner, recombine a fraction of their genome with the genomes of others, merging and unmerging promiscuously. This lateral gene flow renders the notion of independent species among them problematic. Yet for the first three billion years of life organisms were exclusively microbial; today about 90 per cent of earth's biomass is microbial. A different pattern of evolutionary relationship at that time scale and spatial spread probably needs to be figured out. As E. O. Wilson quips, "The bacteria await biologists as the black hole of taxonomy."[10] A new image that does justice to bacteria's pervasive horizontal transfer of genes needs to be found, along with its relationship to the tree of life.

The above instances show that as new knowledge arises from the sophisticated work being done in many fields, Darwin's explanations for certain phenomena need to be corrected or nuanced. Nevertheless, the theory of evolution itself shows a core ability to accommodate such discoveries pertaining to this or that detail while its major insight has remained intact. In its neo-Darwinian form wedded to the science of genetics, it continues to have unparalleled explanatory potential for the origin of species.

Later in this book we will deal with the emergence of the human species, but it is important to underscore here that on many fronts scientific discoveries have strengthened Darwin's insight that human beings, too, have come into being as a result of descent with modification. In the great peroration at the end of *Origin*, Darwin suggested that "Light will be thrown on the origin of man and his history" (488). This was a colossal understatement. Almost immediately his book got people to think about themselves and their origin in disconcerting new ways. Two of his later major works, *The Descent of Man, and Selection in Relation to Sex* (1871) and *The Expression of Emotions in Man and Animals* (1872), sought to demonstrate human evolution in detail. As a biologist Darwin posited that humans were part of the animal world, inheriting from pre-existing ancestors physical and emotional traits as well as social instincts such as cleverness, kindness, and altruism.[11] There was as yet no field work on which to draw; hominid fossils had yet to be discovered; and the science of genetics was still in the future. Hardly imaginable was the now current theory that the demise of the dinosaurs 65 million years ago opened environmental niches into which surviving small mammals expanded, leading eventually to the evolution of the mammalian species *Homo sapiens*. Hence the precise details of Darwin's explanation have not held up. Thanks to decades of work, however, recent discoveries in paleontology, comparative anatomy, developmental biology, and genetics are helping science write solidly-based even if tentative chapters of human evolution that add new specificity to the core picture Darwin painted.

A COSMIC LENS

The theory of the evolution of life traces a unique biological phenomenon on one cosmic body, Earth. This planet does not stand alone, however. It is part of a solar system, which is part of the Milky Way galaxy, which is part of a neighborhood of galaxies, which form a section of the expanding universe. Nesting the 3.5 billion year long story of earthly life within the more than 10 billion years of cosmic development that preceded it highlights the continuity and discontinuity of life on Earth with the rest of the universe. In our day philosophers of science call upon the wider frameworks of cosmology and astronomy, themselves exploding with new discoveries, to arrive at a deeper grasp of evolution's significance.

Inserting biological evolution into the context of cosmic evolution makes life's propensity to create novelty more comprehensible, if still a wonder.[12] Current scientific consensus holds that the universe originated about 13.7 billion years ago in a primordial flaring forth rather inelegantly named the Big Bang. From that explosive instant to this day, the universe continues to expand. In fact it was Edwin Hubble's discovery in the early twentieth century that the galaxies were rushing away from each other that led to the dating of the cosmos' origin. Run the history rearwards and the universe shrinks to a point of startup. This instant can hardly be described in credible fashion. All of the present universe was somehow packed into a point smaller than that of a typical atom, at extremely high temperature. This infinitesimal spark exploded. The event is an unrepeatable instance explainable by no known laws of physics or anything else. The explosion generated all the matter-energy (Einstein showed they are convertible) that make up the world. Cosmologists suspect that almost as soon as the explosion began an initial inflation took place whereby the originating universe jumped in size by an enormous factor. The universe started to expand to staggering, inconceivable distances across intergalactic space.

It is fascinating to reflect on the fact that even in that early phase, the rate of cosmic expansion was calibrated "just right:" too high, and matter-energy would have flown apart and thinned out so fast that no structures could have formed; too low, and the universe would have recollapsed on itself. The proper rate of expansion created the right conditions for galaxies with all their different bodies to form. It is also significant that the exploding matter-energy was not evenly distributed but was "lumpy." At the level of the very small, such proximity allowed the building blocks of matter, simple atoms, to form, each with a nucleus of protons and neutrons ringed by a circle of electrons. At the level of the very large, proximity allowed gravity to pull chunks together, creating heavenly bodies. Matter clumped, then complexified: "Nature aggregates and builds."[13]

Under the attraction of gravity, hydrogen atoms swirled together. The friction of their close encounter ignited local explosions, creating stars. Pulled by the same attractive force, stars swarmed together to form galaxies in spiral, elliptical, or irregular shapes. Stars are furnaces in which heavier elements are forged—carbon, oxygen, iron, sulphur, nitrogen—thus making the universe more complex than it was in the beginning. These elements, as Rolston observes, "are synthesized in proportions that make later planets and life possible."[14] Stars run their course and in their death throes some explode as supernovae, dispersing the heavier elements from their interior throughout space. This material becomes part of the mix when gravity again pulls clouds of matter together into new tight configurations that ignite as new stars. Telescopes today allow us to witness hot regions where young stars are still being formed by this process throughout our own and other galaxies.

Just such a process occurred some 5 billion years ago on one arm of the spiral galaxy we now inhabit and call, as human observers, the Milky Way. Supernova explosions created a thick interstellar cloud of dust and gas. A great clump was pulled together by gravity and reignited to become the star we call our sun. Much of the rest of this thick cloud got concentrated in smaller chunks not big enough to catch fire, forming the planets and asteroids of our solar system, including Earth. This planet was beautifully positioned, close enough to the sun to catch a goodly portion of its rays yet not so close as to become intolerably hot. Through time and a series of cataclysmic and quiet events, the planet acquired water and an atmosphere. Its basic physical characteristics enabled the development of complex chemistries requisite for life. We do not know exactly how life originated. But when this biological amazement finally occurred, it did so not apart from but within the matrix of the cosmos itself. An extraordinary degree of precision or "fine tuning" in the cosmos' basic structures, laws, and properties of matter-energy set up the conditions for life as we know it to begin.

Placing the origin of species within the larger framework of the history of the universe casts an illuminating light on life on Earth in several specific ways.

Location in time
The 3.5 billion year old story of life is a novel but continuous later chapter in the 13.7 billion year history of the cosmos. Carl Sagan memorably used the device of a one-year calendar as a way to conceptualize the almost unimaginable sweep of time of cosmic history.[15] By his estimate, if the Big Bang occurred on January 1, then the Milky Way galaxy formed in early May, and our solar system in early September. Later that month the first stirrings of primitive life forms began on planet Earth. By November, multi-cellular organisms with nuclei came into being. The days of December saw a sequence of species emerge, ranging from worms, fish, land plants, insects, and amphibians, to trees, dinosaurs, mammals, birds, and flowers. By the end of the month hominids had emerged. Modern *Homo sapiens* appeared sometime in the last minutes before midnight on December 31. Life on Earth happens within the time of the universe.

Location in space

The observable universe is incomprehensibly large. Extrapolating from a series of observations made by the Hubble telescope that produced the famously beautiful Hubble Deep Field images, scientists figure there are at least 100 billion galaxies, each comprised of billions of stars, and no one knows how many moons and planets, all of this visible and audible matter being only a fraction of the matter and energy in the universe. Earth is a medium-sized planet orbiting a medium-sized star toward the edge of one spiral galaxy. Its amazing explosion of life is a rare, surprising event taking place within the space of the ongoing expansion of this lavish universe. While life may exist elsewhere, the immense evolutionary epic of life on this planet is a unique chapter in cosmic history. It will never be repeated.

Cosmic interrelationship

Life is a dynamic state of matter. The world of matter-energy resulting from 10 billion years of cosmic history, when taken over by the world of life, enters into what science calls a state of information carried in the genes. This is information in the organism about how to grow itself (the development of the embryo), how to regenerate itself (metabolism), and how to replicate itself (reproduction). Imbued with information, matter proves to be plastic, flexible, remarkable in its zest to self-organize into complex structures and its capacity to evolve. Yet it is still the same basic material that was formed in galactic events. Scientist and theologian Arthur Peacocke traces the connection in a lucid manner. What makes the blood of humans and other mammals red? Answer: iron. Where does this iron come from? "Every atom of iron in our blood would not be there had it not been produced in some galactic explosion billions of years ago and eventually condensed to form the iron in the crust of the earth from which we have emerged."[16] Poetically speaking, living creatures are composed of stardust, or in more prosaic terms, leftover products of nuclear explosions. It is life's energized information that makes a transforming difference in this material. In an astonishing way, when living organisms arrive physical nature rachets up to a new level, while species remain connected to the cosmos in the cell of every member.

Dynamism

The adventure of life participates in the ongoing dynamism of the universe itself. Over time new entities, structures, and processes come into being that did not exist before. They change and complexify, to be replaced by yet newer forms. Out of the Big Bang, the energized atoms; out of the atoms the galaxies of stars; out of the stardust, the Earth; out of the minerals and gasses of the Earth, single-celled living organisms; out of the evolutionary life and death of these creatures, an advancing tide of life, fragile but unstoppable, from sea to land and air; from plant to animal life; and very recently from primates to human beings. Once life begins, there is a disposition in biological nature to improvise, to be creative in ways that cannot be foreseen. While the narrative of life is unique, it partakes of the forward drive of the cosmos which has brought forth a suite of fantastic structures ever since the initial flaring forth.

Open-endedness

The story of life's evolution on Earth has an unpredictable, eventful character. In this it partakes of another quality of the universe as a whole, its historical open-endedness. At the instant of the Big Bang, it was not inscribed in the fireball that this galaxy would form here or that solar system take shape there. The circumstances of their formation were genuinely contingent, that is to say, not necessary from any of the forces in play. It is not the case that all conditions were chaotic or that everything happened randomly. The universe is basically ordered, structured by a set of lawlike regularities. Yet the world's development has the character of genuine history where unexpected events create new opportunities for creative advance. Such is the view of twentieth-century science which has brought to an end the mechanistic view of the world associated with Newtonian physics. An openness regarding outcomes is now seen to hold true for events at both very small and very large magnitudes of space as well as for biological events through the long reaches of time.

Very small: quantum mechanics which works at the infinitesimal level of the atom and its subatomic particles has uncovered a realm where time, space, and matter itself behave in ways that have indeterminacy built into them. Statistical probability lends a measure of order to this realm, but precise subatomic events do not seem to occur according to any discernible regularity. For example, while it can be predicted that a certain mass of radioactive uranium will decay within a given time, there is no way to predict which atom will decompose next, or why.[17] Furthermore, as the Heisenberg uncertainty principle asserts, a human observer cannot simultaneously plot both the position and the velocity of a subatomic particle, for by charting one we disturb the other. Does this human inability to nail down and predict subatomic events point to the poor state of our equipment? Or might it rather be due to an ontological indeterminacy in reality itself? Many philosophers of science argue for the latter. Judging from the realm of the very, very small, the fundamental building blocks of the world are neither mechanically pre-programmed nor utterly chaotic, but spontaneous within an orderly system.

Very large: chaos theory has explored a similar open-endedness in certain non-linear, dynamic systems at the macro level, illuminating how in these systems very slight changes in initial conditions ramify upward to produce massive effects.[18] To cite a well-known example from the study of weather systems, a butterfly rapidly moving its wings in Beijing may set up a small air current that interacts with other atmospheric factors, amplifying upward through different levels of intensity to produce a major storm in New York a week later. While the ramifications of change through non-linear systems are regular enough to be traced in mathematical equations, the number of initial conditions that affect each system is so immense and their confluence so unique that human observation will never get a total handle on them. We will never have a completely accurate weather forecast weeks in advance, and this is due not to the limitation of our instruments but to the nature of the weather system itself. Being intrinsically unpredictable in an epistemological sense, non-linear, dynamic systems thus represent a form of "structured randomness" in the orderly functioning of the world.[19] Does this indicate an ontological indeterminacy in the dynamic systems themselves? Many philosophers of science think so.

Over time: just so, evolutionary biology demonstrates that the emergence of life has followed no pre-determined blueprint but is shot through with surprise. Genetic mutations caused by the sun's ultraviolet rays or exposure to chemicals cause variations in the structure and behavior of living organisms. Natural selection favors the ones that adapt best to their environment, as seen in their rate of survival and reproduction. On and on goes this process of a hundred thousand variables, dead ends, and breakthroughs. A favorite mind game invites us to turn back the clock to before the appearance of life on Earth and then let it tick away again. Would the envelope of life appear as it does now? Scientists are virtually unanimous in saying "no," so multiple and diverse, so genuinely open-ended and historical, are the factors that have combined to produce our planet's inhabitants (that asteroid 65 million years ago, for example). Intelligent animal life might develop, for we see in retrospect that the matter of the universe has the potential to evolve into complex structures (brains) from which consciousness emerges. But life on Earth would be a community with a different genetic history, and likely a different physical appearance.

Taken together, scientific understandings of the indeterminism of physical systems at the quantum level, the unpredictability of chaotic systems at the macro level, and the long-term random emergence of new forms through the evolutionary process itself undermine the idea that there is a detailed, unfolding plan according to which the world was designed and now operates. Rather, the stuff of the world has an innate creativity in virtue of which the new continuously emerges through the interplay of law and chance: "there is no detailed blueprint, only a set of laws with an inbuilt facility for making interesting things happen."[20] Genuinely random events intersect with deep-rooted regularities, issuing in new situations which, when regularized, become in turn the basis for new unforeseen events. The world has developed unpredictably, even if in retrospect we can spy an overall direction toward greater complexity.

Reading *On the Origin of Species* within the wider framework of the history of the universe as we know it illuminates what a tremendously special phenomenon biological life is, while at the same time highlighting that its evolution is continuous with forces at work within cosmic time and space. Life on Earth arises in tune with dynamics present throughout the universe.

AN ECOLOGICAL LENS

For over a century the connection of biological evolution to Spencer's concept of the "survival of the fittest" along with its social misuse led to the stereotype that evolution consists of nothing more than brutal competition among individuals. It is a dog-eat-dog world, and in the great battle of life only the rugged and the fierce win the day. More recently, the burgeoning field of ecological science has led to a more subtle assessment, emphasizing the interdependence of species in local habitats. Life on Earth does not exist generically, but "as swarms of species of limited geographical distribution."[21] Around the planet species form diverse communities of life, constituted by relations among organisms to each other and to their environment. The settings differ greatly, from deep caves to woodlands to high arctic tundra. Every habitat harbors a unique combination of plants and animals, soil and bacteria, linked in little food chains. These biological ecosystems, in turn, are incorporated into larger inorganic systems of the flow of energy and matter through the seas, in the atmosphere, and over the land.

To cite one example, in the Black Forest of southern Germany the physical base of granitic soil, hillsides laced with small streams, and pools of fresh water in valleys support a large growth of fir trees. These nourish the moth larvae that feed the songbirds which the goshawk swoops down to eat. Along with asters that support insects also eaten by the birds, and shrews and small mice eaten by the raptor, all the organisms of this particular ecosystem are tightly bound to one another. "Energy is carried as in a leaky bucket from one species to another through the food webs of organisms,"[22] returned to the soil and water, and back to the living creatures in unending cycles. All are indissolubly bound in this one place. In Wendell Berry's eloquent words, "They die into each other's life, and live into each other's death ... and this exchange goes on and on, round and round, the Wheel of Life rising out of the soil, descending into it, through the bodies of creatures."[23]

Within different ecosystems, each individual species is exquisitely adapted by the evolutionary process to interact with others in its vicinity. In the Pacific northwest of the American continent, for example:

> An unborn grizzly bear sleeps in her mother's womb. Even there in the dark with her eyes closed, this bear is related to the outside world. She will not have to develop a taste for blueberries or for Chinook salmon. When her tongue first mashes the juice of the blackberry its delight will be immediate. No prolonged period of learning will be needed for the difficult task of snaring a spawning salmon. In the very shape of her claws is the musculature, anatomy, and leap of the Chinook. The face of the

bear, the size of her arm, the structure of her eyes, the thickness of her fur – these are dimensions of her temperate forest community. The bear herself is meaningless outside this enveloping web of relations.[24]

Rather than existing as independent operators, all organisms live in intricate systems consisting of many such dynamic interchanges. Each ecosystem is unique. Each has intrinsic value as one finite instance of life's complex foothold in a particular time and place, with a diversity of species each of which interacts out of its own evolutionary history. And each ecosystem is substantially resilient. If a storm or other disaster breaks up a local site, opportunistic species rush in to colonize the damaged place, setting in motion a succession that over time will circle back to something resembling the original state as seeds from peripheral, undisturbed sites move in. If a major species is eliminated, others multiply to take its place. Too much stress, however, will cause a living ecosystem to erode, its vital interchanges replaced by decay. It is the combination of all diverse ecosystems across the globe, each a little jewel, that has created the great evolving envelope of life on our planet. "This is the assembly of life that took a billion years to evolve," observes E. O. Wilson, with wonder. "It has eaten the storms – folded them into its genes – and created the world that created us. It holds the world steady."[25]

Read with such contemporary awareness, the ecological sensibility that pervades *On the Origin of Species* rises into view. It is a revelation. Darwin saw nothing if not a profound interrelatedness among organisms in every locale, along with the unity of all life throughout time and space. As the reader can tell from having followed the whole long argument, while natural selection works on individual variations, these always exist in the context of community. The energetic mutual relations of organisms in a particular place form the context in which natural selection occurs. "Multidimensional biological interactions" is as good a description as any for what the metaphor "struggle for existence" entails. No evolution would happen at all without the reciprocal, give-and-take relations among creatures, "the relation of organism with organism being, as I have often remarked, the most important of all relations" (350). In this process, Darwin's world is as full of affinities and mutual dependencies as of competition and death.

Once this relational framework is recognized, an ecological drumbeat can be heard throughout the book. *Origin*'s very first sentence speaks of "the geologic relations of the present to the past inhabitants" of South America, meaning the close structural connection between the dead creatures whose fossils lie deep in the earth and creatures who move about alive on the surface. On page 3 we hear of "the mutual affinities of organic beings;" page 4 criticizes those who have no explanation for "the coadaptations of organic beings to each other and to their physical conditions of life;" page 6 acknowledges "the mutual relations of all the beings which live around us." And this is just the "Introduction." Each chapter repeats the theme in a different key. Looking at the woodpecker, mistletoe, and other beautiful creatures with which the world is filled, *Origin* asks, "How have all those exquisite adaptations of one part of the organisation to another part, and to the conditions of life, and of one distinct organic being to another being been perfected?" (60). At the very center of the presentation on natural selection Darwin muses:

> Slow though the process of selection may be, if feeble man can do much by his powers of artificial selection, I can see no limit to the amount of change, to the beauty and infinite complexity of the coadaptations between all organic beings, one with another and with their physical conditions of life, which may be effected in the long course of time by nature's power of selection. (109)

Evolution is a relational process. The sound of mutual relationship is so pervasively present in the text one might easily miss it. The beat goes steadily on, until the book closes with its vision of the entangled bank, its elaborate forms of plants, birds, insects, and worms "so different from each other, and dependent on each other in so complex a manner" (389). Darwin's view of life is bent on community. The struggle for life is contextual, each species taking from and benefitting others. There would be no evolution without species constantly interrelating with each other in their particular environment.

Along these lines, it is significant that the grand summarizing symbol in *Origin* is not a great chain of being with its hierarchical ranking, nor a ladder with its stratified ordering of rungs, nor ascending steps of progress, all of which fix organisms in individualized permanent positions. Fully cognizant of life's inextricable web of affinities, Darwin proposes instead the tree of life, a branching, interconnected system with kinship in every pore. This gorgeous symbol illuminates an ecological truth: "Let it be borne in mind how infinitely complex and close-fitting are the mutual relations of all organic beings to each other and to their physical conditions of life" (80). This is true not just here or there but everywhere; not just now and then but always. That all species are related in the flow of life and death is a keystone of evolutionary theory. The grandeur displayed in this view of life is ecological in character. Reading *Origin* with an ecologically open eye allows its deep appreciation for the interrelatedness of life to emerge, offsetting the stereotype that life consists of nothing more than brutal competition.

The point of this chapter has been to update *Origin*'s theory of evolution, tracing the ways its insight has been nuanced in the light of scientific advance. For all the subsequent shifts, the book's one long argument remains standing on strong legs: life evolves. Once there were no heartbeats on this earth. Once there was no smelling, swimming, flying, running, singing, purring, barking, mating, preying, or hiding. Once there were no eggs hatching, no mothers nursing young. Once there was no pleasure, no pain, no hunger, no satiety, no birth, no sex, no death.[26] The theory of evolution offers a natural explanation of how all this grandeur came to be. It tells a dramatic tale filled with struggle and serendipity, tragedy and surprise, the concrete end of which is not yet known.

What a great scientific advance like this offers is certainly not an answer to every question but a suite of insights that lead to new ideas and new questions which deserve attention. For religious communities who believe that the living world is God's good creation, the theory of evolution is theologically consequential. How shall we speak of the overflowing love of the creating, redeeming, re-creating God of life in view of evolution? How shall we act toward the natural world in a way coherent with this understanding? Ask the beasts, the birds, the plants, the fish and they will tell you, counsels the book of Job (12.7). As they interact on the entangled bank, their story of diversifying descent is a spellbinding drama, beyond what the author of Job could possibly have imagined. For the sake of the intelligibility of belief in our day as well as a basis for right moral action, it is essential, as John Haught argues, that a "Christian theology of evolution locate this drama within the very heart of God.[27] To this task we now turn.

5

The Dwelling Place of God

How can reason tolerate that the divine majesty is so small that it can be substantially present in a grain, on a grain, over a grain, through a grain, within and without ... entirely in each grain, no matter how numerous these grains may be? And how can reason tolerate that the same majesty is so large that neither this world nor a thousand worlds can encompass it and say 'behold, there it is'? Yet, though it can be encompassed nowhere and by no one, God's divine essence encompasses all things and dwells in all.

Martin Luther

"WE ARE FECUND AND EXUBERANTLY ALIVE"

The first thing the plants and animals teach us when we ask about their religious meaning is that they are created:

Ask the beasts and they will teach you ...
the hand of the Lord has done this.
In his hand is the life of every living thing.

(Job 12.7, 9-10)

This text's poetic way of speaking about creation makes clear that organisms subsist on a foundation other than themselves. Their vitality is a gift from the generous goodness of God whose "hand" gives and supports all life. At their very core creatures live in reliance on this giving relationship, without which they would not exist at all. Clearly, the view that species are created is a religious affirmation, one not afforded by the workings of natural science. It is the expression of a basic trust that the universe has an ultimately transcendent origin, support and goal which renders it profoundly meaningful. When the beasts teach that they are created, they mean that the dynamic presence and activity of the living God undergirds, enfolds, and bears up all evolutionary process.

While creation often gets pinned to the past beginning of things, and rightly so, it is a doctrine with unsuspected depths. Classical theology speaks of creation in three senses as *creatio originalis*, *creatio continuo*, *creatio nova*, that is, original creation in the beginning, continuous creation in the present here and now, and new creation at the redeemed end-time.

At the outset, being created means that plants and animals receive their life as a gift and exist in utter reliance on that gift. Owing their existence to God is the very core of the doctrine. In ultimate terms they do not bring themselves into being nor does their existence explain itself. Their very being here at all relies on the overflowing generosity of the incomprehensible Creator who freely shares life with the world: "In the beginning God created the heavens and the earth" (Gen. 1.1).

Beyond their fundamental origin, being created also means that plants and animals continue to be held in life and empowered to act at every moment by the Giver of the gift. Without this sustaining power they would sink back into nothingness. A beautiful metaphor from a 20th-century philosopher expresses this insight: the Creator "makes all things and keeps them in existence from moment to moment, not like a sculptor who makes a statue and leaves it alone, but like a singer who keeps her song in existence at all times."[1] There is an ongoing relationship involved. An unbroken flow of divine goodness sustains the existence of the universe in every instant, while creatures exist with an absolute reliance on this life-giving power for their own being and action. Divine creativity is active here, now, in the next minute, or there would be no world at all. Theology traditionally speaks about this music in language of the Spirit, the personal presence of the transcendent God: "The Spirit of the Lord has filled the world, and that which holds all things together knows what is said" (Wis. 1.7).

The evolving history of life is still underway. In and through the suffering and death of billions of creatures new forms continue to emerge, and what lies ahead is not yet known. The ever-creating God of life, source of endless possibilities, continues to draw the world to an unpredictable future, pervaded by a radical promise: at the ultimate end of time, the Creator and Sustainer of all will not abandon creation but will transform it in an unimaginable way in new communion with divine life. Being created means that living creatures are the bearers of a great and hopeful promise: "Behold, I make all things new" (Rev. 21.5).

In due course we will examine all three dimensions of creation in some detail. In truth, they cannot be neatly separated, for all are intertwined actions of the love of the creating, redeeming, re-creating holy mystery whom people call God. Instead of going in chronological order and starting with original creation, however, it proves fruitful for our purposes of dialoguing with Darwin to enter into the subject through the door of *creatio continuo*, the ongoing creative relation of God to the world and the world to God in the midst of time. Our exploration starts close to the ground, with the entangled bank now existing. We inquire of its animals, birds, plants, and fish the meaning of their testimony that they are the work of God's hands.

For creation to be continuously happening, the Creator must be continuously present and active. With an eye on the natural world this chapter discusses the first of these aspects, divine presence. This entails starting our dialogue by attending to the Spirit of God, dynamic ground and bearer of all evolution. After noting influential obstacles that have led to this presence being downplayed or even forgotten, we explore select theological, biblical, and philosophical ways of retrieving this presence in thought and imagination. The whole discussion plumbs the classical doctrine of divine omnipresence which holds true across the universe in general to find its connection to the evolving world in particular. The chapter concludes by drawing out the logic of divine presence for the entangled bank. If the Giver of life be always and everywhere present, then the world of life is not devoid of blessing but is itself a dwelling place of God. Once we have established the Spirit of God's continuous presence to all manner of creatures in, with, and under their own naturalness, the following chapter will examine how the ever-present Creator Spirit acts by gifting the natural world with its own operational autonomy. Taken together these insights into divine presence and action begin to fill out what is meant by continuous creation.

OBSTACLES

In the early Christian centuries it was understood that theology was something like a three-legged stool held up by interlocking considerations of God, the human race, and the natural world. As theology developed in the West, however, the importance of the natural world, while never denied, received little sustained attention. Focus turned with much more vigor to the anguish of the human dilemma, to Christ's redemption of sinful human beings, and to the moral demands entailed in living a saved life. Such concentration tended to overshadow the importance of creation as a religious idea. A brief sketch of several factors which contributed to this erasure shows how deep the roots of the problem go.

Contemporary critics point to the thought pattern of hierarchical dualism articulated by Hellenistic philosophy which divides reality into two separate spheres, spirit and matter, and ranks them as being of greater and lesser value. On the one hand, spirit is a transcendent principle expressed in act, autonomy, reason, the soul, whatever is light, permanent, infinite. Matter, on the other hand, is an inferior principle manifested in passivity, dependence, emotions, the body, whatever is dark, transitory, finite. In some conceptual frameworks spirit and matter thus distinguished existed in a harmonious tension of opposites. In the Gnostic forms in which it affected Christian theology, however, these two spheres of existence came to be seen as polar opposites and their differences were maximized. For the human person as an individual, this meant that the body was less valuable than the soul, which was prized as closer to the sphere of the divine and meant to rule over the recalcitrant flesh. Such dualism also elevated human beings as a whole, blessed with rational souls, over Earth's other living creatures which were allied with matter, and thus of lesser worth. The spirituality typically associated with this thought pattern was propelled by the metaphor of ascent: to be holy a person must flee the material world and rise to the spiritual sphere where the light of divinity dwells. One must turn away from nature in order to have communion with God.

Feminist critics note the long-standing connections between this construal and the structures of patriarchy based on the alleged superiority of man, identified with spirit and reason, over woman, identified with body and passions. The hierarchical dualism of spirit over matter got translated into the social hierarchy of men over women, who by definition exist with an inferiority for which there is no remedy. At the same time, the fact that women are birthgivers who bring forth new life out of their own bodies put them in close symbolic correlation with the natural world, with Mother Earth, who likewise brings forth life from within her dark recesses. In the dualistic framework, the physically fecund powers of both women and the earth are ontologically inferior to the rational mind. They are meant to serve men's needs for progeny and life-maintaining skills while men must struggle against the flesh, change, and death which they represent. The resulting worldview subordinates both women and earth to men's control, which can turn violent and exploitative with little compunction.[2]

At the dawn of the modern era the ancient tree of hierarchical dualism received a new layer of foliage in the philosophy of René Descartes. As he saw it, the world is divided into the human rational mind which knows (*res cogitans*) and all other things which are the object of knowledge (*res extensa*). As part of the outer, objective world nature is fundamentally inert and passive. Lacking the inner mental world which constitutes the essential self, its obvious liveliness is simply a thin veneer laid over a mechanized base. Man's rational mind is meant to probe and manipulate its components without compunction. One shudders to watch this view justify the dissection of live animals on the grounds that since they are just "organic machines," mere automata, they do not feel pain.[3]

Hellenistic dualism, patriarchal androcentrism, Cartesian dualism: in themselves these are philosophical systems. But when their patterns of thought were brought to bear on theology, they led to religious reflection that by and large devalued the earth as a decaying present reality over against heaven, an eternal spiritual reality. Even when the natural world was granted a certain value, this was based on its usefulness for human beings, not its own merit. These frameworks of thought have little room for the natural world's intrinsic worth.

It was not only its philosophical partners that led theology to this diminished place; theology made its own novel contributions. In retrospect, one of the most far-reaching was the natural-supernatural distinction. This idea was introduced in the 13th century to distinguish the realm of human nature, human beings taken as simply created, from that of grace, the gift of God leading to salvation. The distinction was intended to protect divine freedom insofar as human beings could exist by nature even without grace. Hence God's gracious gift is unowed. No one deserves or can earn it; it is given freely thanks to God's overflowing merciful goodness. This is, of course, true, and its emphasis rightly forestalls all efforts at human self-justification.

The difficulty arose, however, when emphasis on God's free gift of grace led indirectly to neglect of divine initiative on the other, so-called natural side of the ledger. Theology began to draw the implication that non-graced nature, both human and non-human, had little to do with divine graciousness. The natural world in particular, not caught up in the history of sin and grace, had a simply natural character. Consequently, in David Burrell's astute insight, late medieval theology drew the implication that the natural world is not a "gift" but simply a "given."[4] Once started down this road, a conceptual device originally intended to help theology articulate an important point about divine freedom ended up dimming the meaning of the first article of the creed. The natural-supernatural distinction "unwittingly augmented the tendency to 'naturalize' the created universe and so further obscured the theological import of the Christian profession of faith in God the creator."[5] While it is proper for the natural sciences to approach the universe purely as a given, something simply there without religious connotation, theology's reflection on God who is "maker of heaven and earth" would seem to require a different assumption at the outset. On the strength of the natural-supernatural distinction, however, the truth that Earth exists thanks to the gracious act of creation, its very continuous existence a gift freely bestowed, slipped into the zone of forgetting. Nature was natural, not supernatural. Compared to the drama of redemption, it held little interest. The topic of creation came increasingly to be treated under philosophy, the work of reason, while redemption with all its revealed ramifications absorbed the attention of theology. The relation between the two, nature and grace, creation and redemption, became mostly extrinsic.[6] As Joseph Sittler observes, this "doctrinal cleavage, particularly fateful in western Christendom, has been an element in the inability of the church to relate the powers of grace to the vitalities and processes of nature."[7]

Modern biblical scholarship added its own problematic deterrent to "asking the beasts" by crafting a distinction between nature and history and contrasting the two as a template for discerning the true God of Israel.

Against the background of nature deities prevalent in the ancient world, the true Holy One of the covenant was the God who acts in history. The chosen people were enslaved; then by divine action they were freed, brought to Sinai, and led into a covenanted life in a land flowing with milk and honey. The God of the Exodus continued to act in historic events, mercifully leading them on through the ascent of kings, invasions, exile, and return, while prophets interpreted these events in light of the nation's fidelity or infidelity. In this tradition time was linear, moving on to a future promised but unknown. By contrast, the religions with nature deities knew only the cyclic return of season after season, going nowhere. History, not nature, was the metier of divine revelation. Given this idea, biblical scholarship interpreted the Genesis creation stories not only as chronologically later in composition than the Exodus stories, but also as secondary in importance to the redemptive narrative revealed in the arc of history.

Singly and in combination, these intellectual frameworks are among the formidable influences that have hindered Western theology from taking the natural world seriously as a subject worthy of religious interest. Astute criticisms of each are rife in contemporary scholarship. Alert to their undertow, our dialogue with the evolving world of life now under threat seeks a different starting point. The presence of the Spirit of God throughout the world in the act of continuous creation opens one such door.

LIFE AND LOVE: A TRINITARIAN FRAMEWORK

Continuous creation affirms that rather than retiring after bringing the world into existence at some original instant, the Creator keeps on sustaining the world in its being and becoming at every moment. The Nicene creed can be read as implying this truth. The first article of the creed confesses belief in one God, "maker of heaven and earth, and of all things visible and invisible," thus affirming God as Creator of all things in the beginning. But in an interesting way the third article of the creed revisits this subject, confessing belief in the Holy Spirit "the Lord and Giver of life." The Latin word translated as Giver of life, *vivificantem*, shines a spotlight on the dynamism that is intended. The Spirit is the vivifier, the one who quickens, animates, stirs, enlivens, gives life even now while engendering the life of the world to come. Behind this creedal snapshot of the Spirit is a rich biblical and theological tradition.

The Hebrew scriptures have a tremendously powerful sense of the one God who is transcendent beyond all imagination. To protect this holy otherness when speaking of God's active engagement with the world, biblical writers often employ figures such as God's word, wisdom, glory, voice, angel, or spirit, which at one and the same time evoke divine nearness while preserving a sense of the ineffable. As one such signifier, spirit is most often connected with divine presence that gives life. From the beginning when the Spirit of God moves over the primeval waters as the world comes into being (Gen. 1.2), to every instant since then when the Spirit is sent forth to renew the face of the earth (Ps. 104.30); from the natural world which is filled with the Spirit of the Lord (Wis. 1.7), to the world of human beings where the Spirit enlightens, imparts wisdom, creates a new heart, emboldens right speech, inspires prophets, advocates for justice, comforts, builds community, and strengthens love, the gift of life keeps pouring out. Psalm 139 praises this life-giving omnipresence in dramatic language:

> Where can I go from your spirit? Or where can I flee from your presence?
> If I ascend to heaven, you are there; if I make my bed in Sheol, you are there.
> If I take the wings of the morning or settle at the farthest limits of the sea,
> even there your hand shall lead me, and your right hand shall hold me fast.
>
> (Ps. 139.7-10)

Up to the bright skies, down to the underworld of the dead; east to the sunrise, westward to the sunset over the Mediterranean Sea: no matter where the psalmist might roam, God's living spirit, equated with divine personal presence and supporting hand, will be there. Filling the world, the Spirit is the dynamic vitality that gives existence to every single thing, calling it forth and holding it fast.

To speak this way is to highlight divine immanence. Biblical and theological language about the Spirit is marked by the same transcendence that characterizes the whole concept of God, but is distinct in bringing the beyondness of holy Mystery into intimate contact with the world. Such language refers to nothing less than the mystery of God's personal engagement with the world, human, planetary, and cosmic, from the beginning, throughout history, and to the end, calling forth life and freedom. As Walter Kasper eloquently writes, wherever "life breaks forth and comes into being; everywhere that new life as it were seethes and bubbles and even, in the form of hope, everywhere that life is violently devastated, throttled, gagged, and slain; wherever true life exists, there the Spirit of God is at work."[8]

Insofar as this continuously creative activity of giving life is the work of a generous and compassionate Giver, we are on the right track to add that language about the Spirit refers to God's gracious love in person moving with power at all times and in all places. The book of Wisdom illuminates the interlacing of love with the life-giving Spirit in simple, eloquent words. Having reflected on the living God's ways with the people of Israel, the sage expands the horizon to include all of nature:

> For you love all things that exist,
> and detest none of the things that you have made,
> for you would not have made anything if you had hated it.
> How would anything have endured if you had not willed it?
> Or how would anything not called forth by you have been preserved?
> You spare all things, for they are yours, O Lord, you who love the living.
> For your imperishable spirit is in all things.
>
> (Wis. 11.24–12.1)

Read these words with your image of the entangled bank in mind, and already it slips out of being relegated to a natural as opposed to a supernatural realm. The imperishable Spirit in all things is God's own spirit, the spirit of love, calling forth, preserving, and opening to the future everything that exists. Consequently, as Kasper notes, the natural world "is already always more than pure nature. Through the presence and action of the Holy Spirit creation already always has a supernatural finality and character."[9]

This sense of the transcendent God's presence in the Spirit is carried over into the New Testament, but now with a noticeable difference. "The grace of our Lord Jesus Christ, the love of God, and the fellowship of the Holy Spirit be with you all" (2 Cor. 13.14). This short phrase, written only three decades after the death and resurrection of Jesus, provides a key to the early Christians' experience of the living God. Already they had to talk of the Holy One in a threefold manner in order to do justice to what they had experienced and knew deep in their souls: the ineffable God, who had created the world out of love and was infinitely beyond them, had encountered them personally in the historical life, death, and resurrection of Jesus Christ and his mission, and was profoundly present and active among them in a new way in the outpouring of the Spirit that formed their community. There was as yet no formal doctrine of the Trinity. But their faith experience required this kind of threefold language.

One of the earliest attempts to articulate in more systematic fashion the idea that the Creator is a triune God was that of Irenaeus, second-century bishop of Lyon. Psalm 33.6 had long proclaimed, "By the word of the Lord the heavens were made, and all their host by the breath of his mouth" (breath from the Hebrew *ruach*, which also means spirit). Interpreting word and spirit in a distinctive trinitarian framework, Irenaeus came up with a favored metaphor that envisioned the Father creating the world with "his two hands, the Son and the Spirit, the Word and the Wisdom."[10] Later interpreters parceled out the work of creation so that the Son/Word confers rational form or order on the world, and the Spirit/Wisdom animates it with movement and radiance. David Jensen catches how well this image evokes bodily communion, with the Creator drawing near to convey life: "God's activity is accomplished in the touch of two hands,"[11] two graphic ways of reaching out to create the world. Both Word and Wisdom carry out the transcendent God's work in shaping, sustaining, and gracing the world.

Using picturesque nature metaphors, the North African theologian Tertullian, whose work straddled the second and third century, sorted out the trinitarian relationship to the world in a different manner. If God the Father can be likened to the sun, source of light and heat, then Christ is the ray of sunlight streaming to earth (Christ the sunbeam, of the same nature as the sun), and the Spirit is the suntan or sunburn, the spot of warmth where the sun actually arrives and has an effect. Similarly, the triune God can be likened to an upwelling spring of water in the hills, the same water flowing downhill in a stream, and the water in a canal or irrigation ditch where it actually reaches plants and makes them grow. One other metaphor sounds more mellifluous in English: the triune God is like the root, the shoot, and the fruit of a tree, that is, its deep unreachable foundation, its sprouting up into the air, and its burgeoning in flower, fragrance, fruit and seed.[12] Note how the homely metaphors of suntan, irrigation ditch, and fruit point to the effective presence of the divine Spirit at work in the world.

These nature metaphors are trying to express what can never be completely defined, the presence of the incomprehensible mystery of the living God encountered in Jesus Christ and the Spirit unleashed in his company. There is God the unoriginate origin and source of all, who as God comes forth personally in the flesh to be with us in history, and who as God again actually dwells within and has an effect upon the world. In view of the history of salvation, Christians confess belief in one God beyond the world (transcendent), with the world in the flesh (incarnate), and within the world bringing it to a blessed future (immanent). "And it is all one love," as Julian of Norwich so beautifully declares.[13] When the trinitarian God is considered in relation to the world, the Spirit is always God who arrives in every moment, drawing near and passing by, indwelling, gifting, and calling forth with life-giving power.

Trinitarian controversies soon shifted reflection into less picturesque categories, forging doctrine about the co-equal divinity of the three *hypostases* or "persons" (but not persons in the contemporary sense) related in the communion of one divine nature. A milestone was reached with the fourth-century Nicene-Constantinopolitan creed. There is one God, the Father Almighty who creates heaven and earth; there is one God, the only-begotten Son who to redeem the world became incarnate and lived a human life unto death and resurrection; there is one God, the Holy Spirit, Lord and Giver of life, who vivifies the world now and into the future. Subsequent theological developments clarified that these three articles of the creed do not merely reflect a Christian experience of God, but correspond to the inner relational tri-unity of God's own self. The way the Christian community came to know the one God through the Word's becoming flesh and the Spirit's enlivening the community of the faithful corresponds essentially to a triune relationality within God's own being.

To recap a long history with attention to the Spirit: Western theology came to perceive the Holy Spirit as gracious love, proceeding from the Father and the Son and linking them in mutual and reciprocal unity, the way the bond of human love unites lover and beloved, to use one of Augustine's eloquent analogies.[14] Hewing to the original form of the Nicene Creed, Eastern theology of the Orthodox churches envisions this love proceeding from the Father alone, not by way of generation like the Son but as the breath of his mouth which accompanies the word and reveals its efficacy, the three "persons" joined in a *perichoresis* or circling movement of life, like a divine round dance. In both traditions the self-communicating love of the trinitarian God in the inner divine life itself (*ad intra*) and in the action of God in the world (*ad extra*) is spoken of in the language of Spirit. This is divine love on the move, going forth with vital power: "the love of God has been poured into our hearts by the Holy Spirit given to us" (Rom. 5.5). Reflecting on the impulse toward the beloved which love causes in the will, Aquinas finds it suitable "that God proceeding by way of love" be called spirit, indeed Holy Spirit, because love implies a kind of moving force, a compelling energy, akin to what we mean by spirit.[15] The important point to keep in mind is that in this context love refers not to something God does or an affection God entertains, but to who God is, graciousness in person. In formal terms the Spirit is God who is love proceeding in person.

In working on this subject I have found this trinitarian framework to be of utmost importance. It secures the fact that language about the Spirit is not about some lesser being or weaker intermediary, but is referring without dilution to the incomprehensible holy mystery of God's own personal being. The Giver of life is not a diminutive or insubstantial godling, a shadowy or faceless third hypostasis, but truly God who is "adored and glorified" along with the Father and the Son, as the creedal symbol of faith confesses. In sum:

> Speaking about the Spirit signifies the presence of the living God active in this historical world. The Spirit is God who actually arrives in every moment, God drawing near and passing by in vivifying power in the midst of historical struggle. So profoundly is this the case that whenever people speak in a generic way of "God," of their experience of God or of God's doing something in the world, more often than not they are referring to the Spirit, if a triune prism be introduced.[16]

The doctrine of continuous creation has the closest possible connection with pneumatology understood in this manner. The stunning world opened up to our wonder by evolutionary biology and ravaged by our consumerist practices calls for attending to the presence of the Giver of life not at a distance, presiding beyond the apex of a pyramid of greater and lesser beings, but within and around the emerging, struggling, living, dying, and evolving circle of life.

POETIC BIBLICAL IMAGES

To grasp the breadth and depth of the Spirit of God's presence in the world, theological reflection is not well served by our popular culture's ordinary image of God as simply a single, anthropomorphic male authority figure in the sky. This stereotype renders God an all-controlling, imperialistic, distant superbeing who starts up the world and intervenes now and then to bring about desired effects. Such an image is "too small" to go the distance for the beasts, let alone for ineffable truth of holy Mystery, and needs expanding toward something ever greater.[17] Toward that end, key biblical images for the active presence of the Spirit are an important resource. Powerful natural forces like blowing wind, flowing water, and blazing fire expand the notion of divine presence beyond analogy with a human person. None of these forces has a definite, stable shape. They can surround and pervade other things without losing their own character; their presence is known by the changes they bring about. Not that the Spirit of God is impersonal. But compared with anthropomorphic images drawn from human beings who are physically limited in time and place, these natural phenomena seem particularly suited to draw out the surging creative energy which religious language seeks to express. So too with the figure of the bird, brooding and flying free, an animal metaphor for divine creative presence. When scripture does draw from the human species, the great figure of holy Wisdom steps forth, a personal expression of the Creator's active presence fashioning and enlivening the world. These biblical ways of alluding to the Spirit's creative presence expand our religious imagination and provide an initial vocabulary with which to explore the hidden depths of the living God in the evolving world.

Wind

In the Hebrew Bible the word for spirit is *ruach*, a word of complex meanings which also translates as wind or breath. *Ruach* can refer to the meteorological movement of air, or again to the life breath of animals or human persons inhaled and exhaled, which breath itself becomes an analogy for the human spirit or self. The common thread is invisible movement that has an effect. Raging in storms or blowing as gentle breezes, winds stir the atmosphere across land and sea, creating weather patterns and dispersing seeds. Breathing air in and out is a sign that persons and other animals are alive; when breath departs, they are dead. Such ordinary observations provide a way of talking about the ineffable movement of God's own Spirit, a divine wind or breath that bears the vital force of life.

The creation account that opens the Bible draws on this metaphor. In the beginning when all is wild and chaotic, the *ruach Elohim*, which can be translated as breath or spirit of God, moves, sweeps, blows like a wind over the face of the waters, and the world begins to take shape (Gen. 1.1-2). The wind metaphor for Spirit has a long subsequent biblical life, perhaps most memorably when it blows over a valley of scattered dry bones so that they reconnect, click together with a rattling noise, get clothed with flesh, and infused with the breath of life, a fabulous prophetic symbol that a vanquished people will have their lives renewed (Ezek. 37.1-14). In the New Testament Jesus explains to Nicodemus that being born anew does not mean literally crawling back into your mother's womb but being reborn in the Spirit, which is like the wind: "The wind blows where it will, and you hear the sound of it, but you do not know where it comes from or where it is going; so it is with everyone born of the Spirit" (Jn 3.8). Untamably free, like the wind, God's own *ruach* has a powerful, rebirthing effect on the human person. The Pentecost story draws on the same imagery to describe the coming of the Spirit upon the 120 or so disciples assembled in the upper room, whipping up their courage: "And suddenly there came from heaven a sound as of the rushing of a mighty wind, and it filled all the house where they were sitting ... and they were all filled with the Holy Spirit and began to speak in other languages" (Acts 2:2, 4).[18]

The Spirit dwells in the world like wind, blowing freely and affecting everything. It cannot be corralled or enclosed, restricted or caged. Both in the natural and the human world, wherever this divine wind blows, something new is stirred up.

Water

It rains and snows from the sky, flows in rivers and small streams, pools in underground aquifers and wells, is channeled through aqueducts and pipes. In salty form it covers three-quarters of Earth's surface. Coursing through the bloodstream, found in every cell, water is essential for every creature's biological life. Deprived of it, they wither; supplied with it, they are refreshed. Biblical use of water as metaphor for the presence of God's own Spirit often appears in connection with the action of outpouring. Speaking hopeful words to the people of Israel suffering in exile, the prophet Isaiah conveys God's promise, encouraging them not to be afraid for:

> I will pour water on the thirsty land,
> and streams on the dry ground.
> I will pour my spirit upon your descendants,
> and my blessing on your offspring.
> They shall spring up like a green tamarisk,
> like willows by flowing streams.
>
> (Isa. 44.3-4)

The metaphor of outpouring water appears regularly in biblical prophets. It is put to good use again in the Pentecost story. Disputing the crowd's criticism that the disciples are drunk (after all, it is only nine o'clock in the morning), Peter declares that their linguistic ability to speak different languages is an instance of what was spoken of through the prophet Joel, whom he then quotes:

> In the last days it will be, God declares,
> that I will pour out my Spirit upon all flesh,
> and your sons and your daughters shall prophesy,
> and your young men shall see visions,
> and your old men shall dream dreams.
> Even upon my slaves, both men and women,
> in those days I will pour out my Spirit;
> and they shall prophesy.
>
> (Acts 2.17-18, citing Joel 2.28-9)

This image of tipping over the amphora, upending the pitcher, decanting the bottle, giving water its free run out of the bucket, letting it flow as rain that comes pouring down, is always associated with the giving of new life, whether to the land, the human person, the community, or the whole world when redemption is finally accomplished. One is reminded of Hildegard of Bingen's guiding image of *viriditas*, greenness, which runs throughout her work expressing the freshness, fertility, and fruitfulness of the life-giving power of the Spirit.

Imagine a large sponge floating in the sea, saturated through and through with ocean water. This is Augustine's way of using water to conjure divine immanence. Not as dynamic an image as water poured out, its very quietness bespeaks the unfathomable depth of God's indwelling. In an imaginative passage in the *Confessions* he first sets before his mind's eye the whole finite creation—earth, sea, and sky, stars, trees, and mortal creatures, everything seen and unseen. Then:

> But Thee, O Lord, I imagined on every part environing and penetrating it, though in every way infinite: as if there were a sea, everywhere and on every side, through unmeasured space, one only boundless sea, and it contained within it some sponge, huge, but bounded; that sponge must needs, in all its parts, be filled with that immeasureable sea: so conceived I Thy creation, itself finite, yet full of Thee, the Infinite...[19]

Like a saturated sponge creation is dripping wet with divine presence, so to speak. Like a soaking ocean, a flowing fountain, an inexhaustible wellspring of sweet water, the life of the Spirit pervades the world. "Through the Holy Spirit," Jürgen Moltmann writes, "God's eternal life brims over, as it were, and its overflowing powers and energies fill the earth."[20] Wherever this divine water flows, life is being refreshed.

Fire
Prized for its gifts of warmth and light but also, like wind and water, at times uncontrollably dangerous, fire symbolizes the presence of the divine in most of the world's religions. Lighting lamps or candles and burning incense is a typical ritual act. Biblical references to fire as symbol of the divine are multivalent, evoking wrath against evil-doers as frequently as life-giving power. Always, however, and especially in key passages, its connection with a special approach of God is unmistakable. A burning bush blazes in the desert but is not consumed: from it Moses hears the compassionate call to deliver people from enslavement, and receives the divine name as pledge that liberation lies ahead (Exod. 3.7-14). After their exodus, fire again signals divine presence as the people are invited into a covenant relationship: "Now Mount Sinai was wrapped up in smoke, because the Lord had descended upon it in fire" (Exod. 19.18). Fire's connection with the Spirit continues explicitly in the New Testament. John who is baptizing with water tells the people that the greater one coming after him "will

baptize you with the Holy Spirit and fire" (Mt. 3.11). Christian religious art has long depicted the scene at Pentecost when, after the great wind rushes through, "divided tongues, as of fire, appeared among them, and a tongue rested on each of them. All of them were filled with the Holy Spirit" (Acts 2.3-4). The fire of the Spirit sets human hearts on fire and inspires boldness, prompting all the disciples to step out and do something new, needed, now.

As in the human world so too in the world of nature: the whole of creation is sparked by the Spirit's presence. In a poetic oracle, Hildegard of Bingen channels the Giver of life:

> I, the highest and fiery power, have kindled every living spark and I have breathed out nothing that can die ... I flame above the beauty of the fields; I shine in the waters; in the sun, the moon and the stars, I burn. And by means of the airy wind, I stir everything into quickness with a certain invisible life which sustains all ... I, the fiery power, lie hidden in these things and they blaze from me.[21]

At the end of his popular book *A Brief History of Time*, which sets out the basic structure of the universe, physicist Stephen Hawking asks a famous question: "What is it that breathes fire into the equations and makes a universe for them to describe?"[22] In the integrity of his adherence to atheism, he leaves the question open. Faith offers a different option, daring to believe that it is God's own Spirit who breathes the fire of life into these equations, indeed, who fires up the equations to begin with. The Spirit dwells within the world like glowing fire; wherever this divine fire burns, creation is sparked into luminous being.

Blowing like wind, flowing like water, flaming like fire, the Spirit of God awakens and enlivens all things. Each of these symbols has a numinous quality that evokes better than more abstract words the presence of the Creator Spirit in the natural world, in plants, animals, and the ecosystems of the earth. The whole complex, material universe is pervaded and signed by the Spirit's graceful vigor, blowing over the void, breathing into the chaos, pouring out, refreshing, quickening, warming, setting ablaze. The import of these images becomes stronger when we recall that speech about the Spirit refers not just to a modality of divine presence, God remaining in heaven, so to speak, while sending the Spirit as a means of outreach to the world. Rather, such language refers to the loving presence of the living God as such, beyond anything we can imagine, creating the power of life in all things.

Bird

One animal more than any other has been used to symbolize the effective presence of the Spirit in the world, namely the bird. To ancient peoples these denizens of the skies seemed closer to the heavenly dwelling place of God, and their freedom of riding the wind and coming to rest on earth came to represent the streaming of divine power to land-bound humans. Christian religious art, ancient and contemporary, focuses in particular on the dove, visually depicting what is recounted in the story of Jesus' baptism: while Jesus was praying the heavens opened, "and the Spirit descended upon him in bodily form like a dove" (Lk. 3.22).

The Bible's use of this symbol of the dove for God's Spirit draws on a rich pre-existing tradition. In Ancient Near East religions the dove was an iconic representation of female deity, whether the Babylonian Ishtar, the Semitic Astarte, Anat in Egypt, or later the Greek Aphrodite, goddess of love. These cooing birds, mentioned in myths, sculpted onto small clay shrines, and tended in cultic towers, came to symbolize the attributes of love, beauty, and fecundity associated with these deities and conveyed as gifts to their devotees. Assimilating this symbolism at various points to its own tradition, the Hebrew scriptures present a constellation of imagery of the bird and her wings. Whether hovering like a nesting mother bird over the egg of primordial chaos at the creation (Gen. 1.2); or sheltering those in difficulty under the protective shadow of her wings (Ps. 17.8, 36.7, 57.1, 61.4, 91.1,4; and Isa. 31.5); or bearing the enslaved up on her great wings toward freedom (Exod. 19.4; Deut. 32.11-12), the approach of God's creative and recreative Spirit is evoked with allusion to this animal and, by association, to the broad tradition of divine female power.

Later streams of Christianity carried forward this interwoven symbol of female bird, powerful divine love, and Holy Spirit. One of the strongest expressions is found in Syriac Christianity where the Spirit's image, consistently linked with that of the brooding or hovering mother bird, brought the idea of divine maternal care to the fore. This local church described the relations of the Spirit to her children in terms of giving birth, nourishing, protecting, comforting, and accompanying into the future, all expressions of divine care intimately present. They noted how in the gospels the Spirit mothers Jesus into life at his conception in Mary's womb, empowers him into mission at his baptism, raises him from the dead, and brings believers to birth out of the watery womb of the baptismal font. In one extant prayer the individual believer meditates:

> As the wings of doves over their nestlings,
> And the mouths of their nestlings toward their mouths,
> So also are the wings of the Spirit over my heart.[23]

In another prayer spoken publicly in the context of liturgy, the Spirit is praised and implored: "The world considers you a merciful mother. Bring with you calm and peace, and spread your wings over our sinful times."[24] The doctrine of the motherhood of the Spirit fostered a spirituality characterized by great warmth, expressed in private and public prayer.

The same maternal imagery of the brooding bird found play in Augustine's interpretation of the Genesis story. Trying to move readers away from the idea that the six days of creation were exactly twenty-four hour days, he argues that God acts not in time as humans do but "by the eternal and unchanging, stable formulae of his Word, coeternal with himself, and by a kind of brooding, if I may so put it, of his equally co-eternal Holy Spirit." Then, alluding to the opening of Genesis where the Spirit of God was *moving/blowing/being borne* over the waters, he riffs on the different nuances of that verb in Greek and Latin, Hebrew and Syriac. Borrowing from the latter, he writes that the verb could mean that the Spirit of God was brooding over the water in the way birds brood over their eggs, "where that warmth of the mother's body in some way also supports the forming of the chicks through a kind of influence of her own kind of love."[25] The notion of a warm maternal bird fostering and cherishing the growth of her young, actually engendering them into existence by the loving power of her own body, provides an apt animal metaphor for the creative work of the Spirit of God, Giver of life.

Wisdom

In the sentence following this reference to the mother bird, Augustine cautions readers not to think of these six days in a literal-minded, childish way, but to grow up into mature appreciation of the way God operates.

Indeed, he says, the reason "why the very Wisdom of God took our weakness upon herself and came to gather the children of Jerusalem under her wings as a hen gathers her chicks was not that we should always remain little children, but that while being babies in malice we should cease to be childish in mind."[26] In the background is the passage in Matthew's gospel where Jesus, facing Jerusalem's rejection, intensely regrets being thwarted in his desire "to gather your children together as a hen gathers her brood under her wings" (Mt. 23.37). Here Augustine identifies the Wisdom of God, vulnerable in solidarity with human weakness, with the person of Jesus Christ, mother hen. In doing so he is embroidering on the wisdom christology of the New Testament, a key ingredient in the early church's move toward the idea of incarnation.[27]

While later Christian theology tended to connect the figure of Wisdom with Jesus Christ, the earlier tradition more often associates her with the world-enlivening presence of the Spirit; recall Irenaeus' construal of Son-Word and Spirit-Wisdom. In the biblical wisdom writings, Holy Wisdom (*hokmah* in Hebrew, *sophia* in Greek) is a female figure of power and might. Assimilated by Israel's sages from surrounding cultures that worshiped female deities in many forms, and fearlessly incorporated into the structure of monotheistic faith as an enriching way of speaking about the one God, this figure enabled Jewish belief to be expressed in a way that matched the religious depth and style of the goddess cult while counteracting its appeal. Biblical wisdom literature's language about Sophia celebrates the one God's gracious goodness in creating and sustaining the world and in electing and saving Israel, and does so by drawing on female imagery of the divine.[28]

Poetic passages of great beauty delineate her cosmic reach. *Proverbs* 8 presents her as present with God at creation, working as a master crafts-person and playfully rejoicing in the result (Prov. 8.22-31). In *Sirach*, Wisdom is connected with swift mobility, nourishing mist, and radiant light (wind, water, fire); she makes a grand proprietary tour of the cosmos, sweeping from the vaults of heaven to the depths of the abyss where she holds sway (Sir. 24.1-6). Recounting the blessings of life, Solomon calls her the "mother"of all these good things, though regrettably he did not always recognize this (7.12); she knows and can teach him the secrets of the natural world because she fashioned them all (Wis. 7.22). Far from being a distant power, her creative agency places her in stunningly intimate contact with everything:

> For wisdom is more mobile than any motion;
> because of her pureness she pervades and penetrates all things.
>
> (Wis. 7.24)

Transcendently beautiful, her radiant goodness is strong enough to defeat even evil:

> She is more beautiful than the sun,
> and excels every constellation of the stars.
> Compared with the light she is found to be superior,
> for it is succeeded by the night.
> But against wisdom evil does not prevail.
>
> (Wis. 7.29-30)

Present everywhere, she is the source of harmonious organization in the world, laying down structures and processes that engender life. All is set and held in order by her encompassing power:

> She reaches mightily from one end of the earth to the other,
> and she orders all things well.
>
> (Wis. 8.1)

There are places in the wisdom literature where Sophia is explicitly identified with God's spirit. "For wisdom is a kindly spirit," literally a people-loving spirit, the book of *Wisdom* declares (Wis 1:6). Shifting the metaphor, this same book affirms that Wisdom herself possesses a spirit described in glorious vocabulary: "There is in her a spirit that is intelligent, holy, unique, manifold, subtle, mobile, clear ...," twenty-one attributes in all, or three times the perfect number seven (Wis. 7.22). Again, wisdom and spirit are held parallel: "Who has learned your counsel, unless you have given wisdom, and sent your holy spirit from on high?" (Wis. 9.17). More persuasive than these directly-stated equivalences between wisdom and spirit, however, is the five-fold metaphor that intrinsically links Wisdom to the mystery of God's own being:

> For she is a breath of the power of God;
> a pure emanation of the glory of the Almighty ...;

For she is a reflection of eternal light;
a flawless mirror of the working of God;
and an image of his goodness.

(7.25-6)

In this pre-trinitarian literature, there is no precise one-on-one correspondence between Wisdom and the Creator Spirit as confessed in the creed. But the similarities of function and relation to divine being are so profound as to allow theology, in the past as now, to adapt wisdom categories for interpretations of the Spirit. The way she moves, breathes, fashions, delights, orders, pervades, and triumphs over evil gives imagery for the transcendent Creator Spirit's gracious goodness in continuously creating the world with a radiance that is the fruit of indescribable love.

Wind, water, fire, bird, holy Wisdom's mobility, beauty, and creative power: these symbols provide guides for how to think about the hidden presence and activity of the Spirit of God in the natural world. This ineffable presence is innermost to creatures, a vital power that enlivens, nurtures, sparks, and fructifies them in every instant. Giving us one more sensory image, Augustine compares the Spirit to the wafted fragrance of a complex perfume, "the sweetness of begetter and begotten pervading all creatures according to their capacity with its vast generosity and fruitfulness."[29] When the Nicene creed calls the Spirit the "Giver of life," *vivificantem*, the Vivifier, it is giving a creedal blessing to the earlier biblical and later doctrinal idea that creation is not just a one-time event in the beginning but entails the presence of the Spirit of God every step of the way. Bounteous love more mobile than any motion, the mystery of the Creator Spirit, utterly transcendent, dwells at the heart of the evolving world in its living and dying, empowering its advance.

THE WISDOM OF PHILOSOPHY: PARTICIPATION

The poetic images of the Bible offer a way for thought and feeling to grasp the expansive presence of God in the world as infinite, life-giving love. Translating this idea into more rational discourse, Thomas Aquinas provides a clear conceptual basis for the same subject. The revelatory story of salvation as experienced in the life of the community supplied Aquinas the Christian with a conviction about the personal presence of the transcendent Creator to all creatures. Aquinas the theologian took philosophical categories and, by extending them in the direction indicated by revelation, made them over into useful intellectual tools for probing this rich faith conviction. The Bible teaches that one God created the heavens and the earth and all that is in them. Given the options in his intellectual landscape, Aquinas thought that the best way to secure the truth that God alone is the source of everything was to regard the one God as the plenitude of being, sheer being itself, while all else participates in being which is given as a gift. His thought provides a philosophical explanation of the immanence of God in the natural world in such wise that nature can never be thought to be godless. When used to interpret continuous creation by the indwelling Spirit, it provides one way to think about nature so that it is connected with the Giver of life from the outset.

The fundamental notion is that God, who cannot be comprehended by any finite idea, simply is. The very nature of what it means for God to be God lies in sheer aliveness, overflowing plenitude, the pure act of being. God's very essence is simply to-be, without origin, limit, or end. Aquinas' Latin is helpful here. One word for being or existence is *ens*, a noun, an entity, something that *has* existence, an actually existing being. Insofar as this limits the idea of the divine to a particular something-or-other, *a being*, this language is not suitable, since the living God is not simply a being among other beings, no matter how transcendent. Another word for being is *esse*, a verb that literally means "to be." Inadequate as all vocabulary is when used of incomprehensible Mystery, *esse* concentrates attention on God as the Verb. To say that God is being in the sense of *esse* means that God is not a noun, not *a* being, not a substance, not a static thing, does not have the property of being, is not in a class described as being at all. Rather, the infinitive form of the verb accentuates the active force of "be-ing," namely "to be," which evokes not a substance but infinite divine aliveness. God is to-be. Think fire.

Given the problems with the idea of being in modern philosophy, some contemporary theologians argue that this philosophical notion serves best when fertilized by theological language about the triune God. If God who is being itself exists as a communion of three "persons" united in love, then this allows being to be understood as a relational category rather than a static, abstract, or impersonal idea. Bringing this insight to bear, reasons Walter Kasper, actually affects a revolution in the understanding of being insofar as now "the meaning of being is to be found not in substance that exists in itself, but in self-communicating love."[30] When we come to understand being as a relational reality even in the field of philosophy, then speaking about God as being itself, fullness of being, pure actuality, connotes the dynamic overflow of life-giving love both within divine communion and in relation to all creation. In a similar manner, Catherine LaCugna mounts a ferocious argument that "God's To-Be is To-Be-in-Relationship."[31] The relational ontology of the trinitarian God is revealed in the economy of redemption in Jesus Christ through the Spirit, from which we are given to know that there is no divine essence which is not at the same time communion. While later interpreters took Aquinas to be discussing the one God in abstract terms apart from revelation, she maintains, a structural comparison between his questions on the one God and the triune God indicates that the supremely existent One is none other than the tripersonal God whose very essence is To-Be-Related. This relational lens allows language about God as the sheer act of being to be filled with the personal richness of the biblical affirmation that "God is love" (1 Jn 4.16). It is this notion of being interpreted as self-giving love beyond imagining that I am using in the discussion below.

Active wellspring of life, God creates the world by giving a share in being to finite creatures in ways appropriate to their own nature. The rigorous distinction between the One who is being itself and all else which receives being intends, rightly, to place the Creator beyond any category commensurate with creatures. At the same time, the act of creation sets up a relation whereby all creatures continuously depend on the originating and sustaining Creator.[32] Exploring this relation, Aquinas asks "whether God is in all things?" His positive answer draws on fire and light in a way that repays careful reading:

> I answer that, God is in all things; not, indeed, as part of their essence, nor as an accident, but as an agent is present to that upon which it works ... Now since God is very being by his own essence, created being must be his proper effect; as to ignite is the proper effect of fire. Now God causes this effect in things not only when they first begin to be, but as long as

they are preserved in being; as light is caused in the air by the sun as long as the air remains illuminated. Therefore as long as a thing has being, God must be present to it, according to its mode of being. But being is innermost in each thing and most fundamentally inherent in all things ... Hence it must be that God is in all things, and innermostly.[33]

What would have been simply a good rational reflection based on cause and effect assumes interpretive power regarding the Spirit by use of the symbol of fire. Just as fire ignites things and sets them on fire, the Spirit of God ignites the world into being. This obviously happens in the beginning but doesn't stop: just as the sun brightens the air all the day long, the presence of the Spirit sustains creatures with the radiance of being as long as they exist. The symbol of fire and its intensification in the shining sun bespeak the *innermost* indwelling of the Spirit throughout the universe, including the creatures of the natural world on planet Earth.

The presence of the Creator Spirit is hidden. It is not discernible by scientific method or instrument, nor can it be thematized as part of any scientific theory. This crucially important point is underscored by Aquinas' answer to the question of "whether God is everywhere." The answer is affirmative. Having created the world, God "is in all things giving them being, power and operation." Then comes an important precision:

God fills every place; not, indeed, like a body, for a body is said to fill place inasmuch as it excludes the co-presence of another body; whereas by God being in a place, others are not thereby excluded from it.[34]

Creation is a *sui generis* relation. God cannot be counted as an additional element in the working of the world. We are not talking about a bodily presence on the matter-energy spectrum that would take up room otherwise occupied. We are not talking about an active presence that would input energy as if it were another causal factor within the created nexus of causes, able to be discovered by science. To the contrary, the very fact that the Creator gives being, power, and operation to everything in every place means that divine presence cannot be ranked with any other factor but undergirds all things as their ineffable Source. Speaking of divine presence in the natural world is faith language, post-suntan language, speech from the irrigation ditch, talk pervaded by sweet fragrance. It is not asserting the presence of the Spirit of God in a quantitative way as if this were one more ingredient added to biological life, but affirming this life in its own created integrity from a theological perspective.

Aquinas understands divine indwelling "in all things" and "everywhere" to entail an interesting mutuality. When bodily things are said to be in another, they are contained by whatever they inhabit. Spiritual things, however, cannot be so easily confined. In particular when we are speaking of God, divine presence spills over beyond the interior of creatures, so to speak, to encompass them on the outside as well. Hence, while "God is in all things," Aquinas argues, it can also be said that "all things are in God," inasmuch as they are "contained" or embraced by a living presence which cannot be limited in any way.[35] Contemporary theology calls this model of the God-world relationship panentheism, from the Greek *pan* (all), *en* (in), and *theos* (God): all-in-God. Simply put, it envisions that the world is indwelt by the presence of the Spirit while at the same time it is encompassed by divine presence which is always and everywhere greater. Rather than conflating God and the world as happens with pantheism, panentheism allows that God who dwells within also infinitely transcends the world at every point. At the same time, it honors the immanence or closeness of God, which is frequently overlooked in unipersonal theism which posits God solely as a transcendent cause. Different from either of those options, panentheism entails a kind of asymmetrical mutual indwelling, not of two equal partners, but of the infinite God who dwells within all things sparking them into being and finite creatures who dwell within the embrace of divine love. In truth, since God in principle does not have any spatial attributes, this is a metaphor whose 'en' expresses the intimacy of relation in an ontological sense.[36] It provides one way of giving intellectual structure to what the apostle Paul communicated to the people of Athens when he preached that God is not far from anyone, for in this ineffable mystery "we live and move and have our being" (Acts 17.28).

The basic view that "God is in all things, and innermostly" and "all things are in God" receives elucidation through Aquinas' use of the notion of participation. For each creature, being created entails "a certain relation to the Creator as to the principle of its being."[37] The idea of participation is one way to spell out this relationship. In creating the world God whose very nature is the plenitude of sheer aliveness gives a share in that vitality in a creaturely way to what is other than Godself. This gifting is free and primordial; without it nothing would exist. For their part, creatures exist by receiving this exuberant divine gift; they are sparked into being as iron glows hot because of fire. "Therefore," writes Aquinas, "all beings apart from God are not their own being, but are beings by participation."[38] Philosophically speaking God *is* being itself, the wellspring of life, while all creatures *have* being, sharing in that livingness through their created acts of existing: "a thing has being by participation."[39]

Parsing this relationship of Creator to creature, Norris Clarke clarifies that participation has three elements: an infinite source, finite things, and a link between the two.[40] The link here is the free giving of being. No finite creature can earn or deserve it, but it is poured out by God through the gratuitous act of creation. One of the words Plato uses to describe participation is *koinonia*, translated from the Greek as fellowship or communion.[41] In Aquinas' fruitful adaptation, the term indeed spells out a profound communion. By virtue of the fact that they exist, creatures participate in ways proper to their own finite nature in the very being of the incomprehensible, self-diffusively good God. In the dynamism of continuous creation, the relation is one of freely gifting and relying on the gift.

The notion of participation affects the understanding of both divine presence and the natural world. On the one hand, there is the intimate and profound presence of the creating Spirit to all individuals, freely igniting them into their own existence. On the other hand, in its own created being and doing, the natural world continuously participates in the livingness of the One who is sheer, exuberant aliveness. It does so, of course, not divinely, but as created, that is, existing and acting according to its own finite nature. In this framework creatures are truly other than God. They exist with their own integrity and are themselves properly agents and causes, in participated finite ways, with a difference from God that is ultimately and essentially good. We encounter that goodness not merely in looking past creatures to their Source, but also in looking at them, in celebrating their intrinsic density and their irreplaceable uniqueness. At the same time, they exist because the loving Giver of life shares the plenitude of being as the grounding source of their existence at every moment. Participation signifies this intimate and profound relationship.

The way this works receives a particularly clear example when the question becomes the goodness of things. God alone is good. But divine goodness is generous, it is generative, it is self-giving. Since "it befits divine goodness that other things should be partakers therein,"[42] every created good is good by participation in the One who is good by nature. It follows that "in the whole sphere of creation there is no good that is not a good participatively."[43] Following this train of thought, Aquinas muses about the teeming diversity of creatures in the world, concluding that it is the excellence of their very difference that expresses divine goodness:

> For God brought things into being in order that his goodness might be communicated to creatures, and be represented by them; and because this goodness could not be adequately represented by one creature alone, he produced many and diverse creatures, that what was wanting to one in the representation of the divine goodness might be supplied by another. For goodness, which in God is simple and uniform, in creatures is manifold and divided. Hence the whole universe together participates in the divine goodness more perfectly, and represents it better than any single creature whatever.[44]

Biodiversity in its own natural way manifests the goodness of God which goes beyond our imagination. Noting how this insight validates the importance of the diversity of species, Denis Edwards notes that "no one creature, not even the human, can image God by itself. Only the diversity of life – huge soaring trees, the community of ants, the flashing colors of the parrot, the beauty of a wildflower along with the human – can give expression to the radical diversity and otherness of the trinitarian God."[45] Indeed for Aquinas, the ontological relationship whereby various creatures participate in the goodness of God is the basis for any speech about transcendent mystery at all, for in knowing the excellence of the world we may speak analogically about the One in whose being it participates.

When the philosophical idea of participation is brought into theological interpretation of the creative presence of the Spirit in the world, it functions to include nature in the orbit of God's gracious love from the outset. The whole natural world exists by participation in the being of God. Subtract participation and what is left is a natural world devoid of the presence of God. But this is an abstraction, a remainder concept. There is no such world in historical existence. All of nature is gifted from the outset with participation in divine *esse*, which in theological parlance is divine life, which revelation discloses is Love. Participating in divine being, the entangled bank thus cannot be looked at from a faith perspective as purely "natural." In its everyday existence simply as created it enjoys an innermost relation to the very livingness of God, a relationship that exists apart from any human act of blessing.

Participation secures the insight that for the world to be created at all, it does not suffice that it be "caused" by a transcendent God who remains, so to speak, at a distance. The world is not simply there as a natural thing, a given. Rather, in its robust naturalness the world exists due to a continuous act of love on the part of the Creator Spirit who shares the gift of being in an ongoing way, indwelling creation, sustaining its life, cherishing its every crevasse. The category of participation provides but one technical way to render the theological claim that the natural world is the dwelling place of God intelligible and, hopefully, unforgettable.

GOD'S DWELLING PLACE

Turning back to the beasts, birds, plants, and fish, it now becomes clear that in the framework of continuous creation the correlative to the vivifying presence of the Spirit of God throughout the natural world is the blest character of that world itself. The inner secret of the entangled bank is the dwelling of God's Spirit within it. Instead of being distant from what is holy, the natural world bears the mark of the sacred, being itself imbued with a spiritual presence. This is not to say it is divine. But unlike gnostic views that disparaged the material world, or the natural-supernatural distinction that divorced it from God's graciousness, the doctrine of continuous creation sees the natural world in its own integrity as the dwelling place of God. The Giver of life creates what is physical—stars, planets, soil, water, air, plants, animals, ecological communities—and moves in these every bit as vigorously as in souls, minds, ideas. Earth is a physical place of extravagant dynamism that bodies forth the gracious presence of God. In its own way it is a sacrament and a revelation.

The developed tradition of sacramental theology teaches that simple material things such as bread and wine, water, oil, the sexual union of marriage, when blessed by the ritual action and prayer of the church, can be bearers of divine grace. This is so, it now becomes clear, because to begin with the whole physical world itself is a primordial sacrament. Pervaded and encircled by the Creator Spirit it "effects by signifying" the subtly active presence of the holy Giver of life. When this presence is channeled through acts of the church, itself constituted as a sacrament of Christ's presence for the world, then the gift of divine grace is conveyed through ecclesially recognized symbols, words, and rituals. In turn, explicit use of these material things by the church in the name of Christ, while specifying the giving of grace here and now, also illuminates the broader "sacramental dynamic"[46] present throughout the cosmos. In view of the eschatological fullness yet to come, John Haught cautions against settling down with an uncritical view of the natural world as sacrament which can get closed in on itself, ignore suffering, and overlook the promise of what is yet to come.[47] Yet the insight that plant and animal species exist by participation in the life-giving power of God does allow for nature's sacramental character to emerge in a critical way, a counterpoint to the forgetful tradition that it is simply a given, not a gift.

A biblically-rooted view of revelation opens another angle of understanding by emphasizing that the natural world can actually teach human beings about God. The heavens tell of the glory of God; day after night after day their speech pours out with knowledge of the Creator whose handiwork they are (Ps. 19.1-4). The book of *Wisdom* makes a strong case that those who do not heed creation's message are foolish and ignorant: "they were unable from the good things that are seen to know the one who exists" (Wis. 13.1). Though it is "the author of beauty" (Wis. 13.3) who creates the magnificence of fire, the might of turbulent water, and the circle of the stars, some wayward folk turn these wonders into idols. The wise, however, allow amazement at the workings of such natural powers to lead them to recognize their source, which is ever greater: "For from the greatness and beauty of creatures the Creator can be seen, so as to be known thereby" (Wis. 13.5).

Pursuing the idea that the cosmos teaches us about its maker, some theologians took to calling nature a book, analogous to scripture. God gave us both books of revelation, and we must learn to read both well in order to glimpse their Author. Augustine exhorted his people vigorously with this metaphor:

Others, in order to find God, will read a book. Well, as a matter of fact there is a certain great big book, the book of created nature. Look carefully at it top and bottom, observe it, read it. God did not make letters of ink for you to recognize him in; he set before your eyes all these things he has made. Why look for a louder voice? Heaven and earth cries out to you, "God made me." You can read what Moses wrote; in order to write it, what did Moses read, a man living in time? Observe heaven and earth in a religious spirit.[48]

What a keen insight, laced with humor. Moses, who was thought to have written the first five books of the Bible, had no books of Moses to read. So what did he do? He read the book of nature. It taught him of the Creator's wisdom and beauty and powerful care.

Pervaded with the mobile motion of divine Spirit, every nook and cranny can disclose the graciousness of the living God. At times the world may be "seared with trade; bleared, smeared with toil," smudged and smelly from human abuse; yet brooded over by the Spirit "with warm breast and with ah! bright wings," in Hopkins' poetic, pulsing words, "the world is charged with the grandeur of God," which will flame out, regardless.[49] This interpretation of the natural world as sacramental and revelatory supports the intense religious experience innumerable people report having when they commune with nature. It also undergirds the anguish that arises in reaction to the destruction of natural places, and sustains the effort to care responsibly for the Earth. Both spiritual and moral responses flow from the understanding of the living world in its givenness, resplendence, fragility, and threatened state as the dwelling place of God.

Pondering the religious value of the biological world, theology in our day needs to make an explicit move to include rather than divide off what has been perceived as merely "natural," i.e., at some remove from God or lacking God's full presence or engagement and therefore of lesser religious or moral significance. The problem is not solved by a simple assertion to the contrary. To be effective a solution must shift the basic structure of thought that desacralized the natural world in the first place. This chapter employs the doctrine of the Trinity as a communion of love, some of the rich biblical images of the Spirit, and the philosophical category of participation to develop the meaning of the doctrine of continuous creation. In so doing it brings back into view the ancient truth of the Creator Spirit's presence pervading creation and the correlative affirmation that the natural world is the dwelling place of God. Building on these insights, the next chapter explores the manner in which the Spirit of God acts in the evolving world.

6

FREE, EMPOWERED CREATION

> ... the cycles of the year and the constellations of the stars,
> the natures of animals and the tempers of wild animals,
> the powers of winds and the thoughts of human beings,
> the varieties of plants and the virtues of roots:
> I learned what is secret and what is manifest,
> for wisdom, the fashioner of all things, taught me.
>
> Wisdom 7.19-22

PARADIGM OF THE LOVER

Up to this point, the theological view being discussed holds true whether the natural world remains static or develops according to the theory of evolution. The Spirit of God is present within the world, continuously sustaining its existence; the natural world is the dwelling place of God's Spirit, able to speak in its own voice about the glory of its Maker. Such understanding is attested to in scripture and is relatively undisputed, though neglected, in the broad catholic tradition. Darwin's entangled bank, however, poses a new question. If indeed its current design is the result of a long history that can be explained by natural laws known to us, how are we to understand not just the presence but the activity of the Giver of life? How does the Vivifier relate creatively to the process of evolution which in scientific terms proceeds according to its own principles? What can the beasts possibly mean when they say "the hand of the Lord has done this" (Job 12.9), given their own unpredictable emergence in the course of evolution? How, in a word, can the evolving world be understood as *God*'s good creation?

Prior to knowledge of evolution, the idea of the Creator went hand-in-glove with the model of God as a monarch ruling his realm. This made a great deal of sense when things could be looked upon as the result of divine design, à la Paley and the watch. Creatures with their myriad features were crafted by divine wisdom, placed in helpful relations to each other, and ruled by providential divine guidance. The whole world reflects the will of the king who holds sway over his kingdom in a direct way. Everything in nature fulfills divine purpose in the way the ruler intends. Reflecting the worldview of their day, a multitude of biblical and classic theological texts express the world's relation to God via this monarchical metaphor.

Difficulty with this picture comes when the theory of evolution makes clear that the world's gorgeous design has not been executed by direct divine agency, so to speak from above, but is the result of innumerable, infinitesimal adaptations of creatures to their environment, from below. The problem gets exacerbated by the fact that the variations on which natural selection works occur randomly. Some adaptations are successful in the current environment and thus filter through to the next generation; some make the organism unfit and so die out; but none are predictable. The absence of direct design, the presence of genuine chance, the enormity of suffering and extinction, and the ambling character of life's emergence over billions of years are hard to reconcile with a simple monarchical idea of the Creator at work. So the question arises: how to understand the presence of the Spirit of God acting continuously to create in the light of evolutionary discoveries about the entangled bank.

Building from a theology of the presence of the Creator Spirit discussed in the previous chapter, the view being proposed here holds that God's creative activity brings into being a universe endowed with the innate capacity to evolve by the operation of its own natural powers, making it a free partner in its own creation. This position differs from deism, where the Creator creates and then leaves the world to its own devices like a clock wound up and left to tick away undisturbed. The difference lies in the presence of the indwelling Spirit of God who continuously empowers and accompanies the evolving world through its history of shaping and breaking apart, birthing and perishing, hitting dead ends and finding new avenues into the future. This position also clearly differs from the kind of monarchical theism where the Creator directly dictates or micro-manages the natural world's every significant move. The difference lies in the idea that the Giver of life freely and generously invests nature with the power to organize itself and emerge into ever-new, more complex forms, and to do so according to its own ways of operating. Far from compelling the world to develop according to a prescribed plan, the Spirit continually calls it forth to a fresh and unexpected future. To be imaginative for a moment, it is as if at the Big Bang the Spirit gave the natural world a push saying, "Go, have an adventure, see what you can become. And I will be with you every step of the way." In more classical language, the Giver of life not only creates and conserves all things, holding them in existence over the abyss of nothingness, but is also the dynamic ground of their becoming, empowering from within their emergence into new complex forms.

This way of understanding God's creative activity in the natural world extends to the natural world what has already been learned about God's gracious ways with human beings in the course of the history of salvation. Toward the end of the New Testament we read the bold statement that "God is love" (1 Jn 4.16), a pithy summary of all that has gone down in the history of revelation up to that point. The phrase testifies to the approach of divine love through the history of Israel and now made newly manifest in Jesus Christ and the ongoing gifts of the Spirit in the church. To develop a theology, as distinct from a philosophy, of God's action in the world, these revelatory events function as an illuminating starting point and ongoing bedrock for reflection. Karl Rahner, for one, has argued that if we see the created world emerging thanks to the self-giving love of God, then the "proper *topos* for achieving an understanding of the immanence of God in the world ... is not a treatise on God worked out in abstract metaphysical terms, but rather the treatise on grace."[1] This is a key move. Methodologically, a theology of divine acts in creation takes its bearings from God's action in Christ through the Spirit, a flashpoint which illuminates how the God of love acts in other contexts.

For all Christian theology, the gospel is good news. The love of God is a saving, healing, restoring power that benefits human beings.

A significant stream of theological interpretation parses this to mean that divine love ultimately enhances the powers of the human person rather being a zero-sum game in which one protagonist's gain is the other's loss. Consider these writers, who share a similar intuition:

 ❦ – Irenaeus penned a classic statement of the God-human relationship with the phrase, *Gloria Dei vivens homo*, "the glory of God is the human being fully alive."[2] God's own honor is at stake in human flourishing, to the point where whenever human beings are violated or their life is drained away, divine glory is dimmed; whenever human beings are quickened to fuller and richer life, divine glory is enhanced. Tying the glory of God so closely to human well-being expresses a precise understanding of the love of the creating, redeeming Mystery as generous, generative, seeking the good of the beloved and having a stake in it.

 ❦ In exploring the relation between divine grace and human freedom, Bernard of Clairvaux reflected a similar understanding of how the Creator's love enhances human autonomy: "What was begun by grace alone, is completed by grace and free choice together, in such a way that they contribute to each new achievement not singly but jointly; not by turns, but simultaneously. It is not as if grace did one half the work and free choice the other; but each does the whole work, according to its own peculiar contribution. Grace does the whole work, and so does free choice – with this one qualification: that whereas the whole is done *in* free choice, so is the whole done *of* grace."[3]

 ❦ The same profound intuition about divine graciousness runs through Karl Rahner's insight that nearness to God and genuine human autonomy grow in direct and not inverse proportion. Put in other words, radical dependence on God and the genuine reality of the creature increase to the same degree.[4] The claim arises in view of Jesus Christ, whom doctrine declares to be truly human as well as truly divine. The deep union of his human nature with divine nature did not render Jesus a robot but constituted him a genuine human being with the integrity of his own freedom. From this central point Rahner reasons that the same dynamic holds true for all human beings. Because grace relates human persons profoundly to the source and goal of their very lives, they become more themselves and can act more freely when they respond to God's gracious self-gift than when they are afar off.

Take these insights about human experience of grace that run through Catholic theology from the second to the twelfth to the twentieth century and extend them to the origin of species. The belief that God is faithful and acts consistently provides a warrant for thinking that as with humans, so too with the natural world from which we have evolved. The gracious God, Spirit proceeding as love in person, is present to bless and enhance natural powers rather than to compete with them. With such a love there can be no anxiety about control.

The one God who creates is also Wisdom made flesh whose self-emptying incarnation into the vagaries of historical life and death reveals the depths of divine love. Could it not be the case that, rather than being uncharacteristic of God's ways, compassionate self-giving love for the liberation of others is what is most typical of God's ways, and therefore also distinguishes divine working in the natural world? In that case we can expect to see not the exercise of controlling power but of divine power as sovereign, cruciform love that empowers others. Given that "Christ is the key to how the Spirit works,"[5] in Kathryn Tanner's felicitous phrase, the same pattern plays out in the activity of the Spirit. The one God who creates and redeems is also the sanctifying Spirit whose self-gift in grace brings healing to sinful hearts and broken situations without violating human freedom. Could it not be that since the Spirit's approach to human beings powerfully invites but never coerces human response, the best way to understand God's action in the evolution of the natural world is by analogy with how divine initiative relates to human freedom? In that case, there is no forcing. Even when the offer of grace is rejected it is not withdrawn; the Spirit graciously continues to invite, prod, push, pull, lure the heart into loving relationship. But the freedom of the creature remains.

There is a lovely logic in the view that a theology of God's ways with creation takes its bearing from the outpouring of divine love on human beings in Christ and the Spirit. The Christ event reveals how God acts, and this rolls over to other contexts. Since gracious divine action expressed in incarnation and the giving of grace reveal the character of God, then holy Mystery who creates, redeems, and sanctifies the world brims over with the most profound respect for creatures. The fear that by drawing near the infinite Creator might crush or destroy the finite creature is unfounded. Rather, seen through the gospel, divine love unfailingly manifests itself not as coercive "power-over" but as "power with" that energizes others. Wolfhart Pannenberg, emblematic of many contemporary theologians, expresses this crucial insight with clarity: "The omnipotence of God can be thought of only as the power of divine love and not as the assertion of a particular authority against all opposition."[6] Active in the world, this loving power accompanies the world as the patient, subtle presence of the gracious Creator who achieves divine purpose through the free play of created processes. In this perspective, the Spirit, more mobile than any motion, blows throughout the world with compassionate love that grants nature its own creativity and humans their own freedom, all the while companioning them through the terror of history toward a new future. Not the monarch but the lover becomes the paradigm.

Love, of course, can be interpreted in myriads of ways; the literature on love could fill whole libraries. Here I single out a homely analogy to clarify the point I am making. Among human persons a mature loving relationship builds up the strength of personal autonomy in those loved, whether they be on an equal footing like spouses or friends or at different stages of life like parents and children, teachers and students. Rather than suppressing the gifts of the other, love brings about their flourishing. Rather than stifling the power to act freely, love promotes its growth. Not all manner of relationships do this. In controlling, manipulative, fearful, narcissistic, and egocentric relations, one party seeks to gain advantage by bending the other to his or her own will. The core integrity of the other is disrespected by ploys that intend to dominate. Mature love moves in the opposite direction. Parents rejoice when their child walks, talks, shows interest in the world, grows into a functioning, contributing person. Teachers rejoice when their students learn skills, pose questions, and begin to think for themselves. Friends, including those partnered in marriage, rejoice when the beloved grows in ability, confidence, power to be uniquely themselves. In our fractured world love is never perfect, always mixed with other forces. On balance, however, its effect is so life-giving because its unifying bond brings about profound growth toward genuine autonomy. In similar yet infinitely dissimilar fashion, creating a universe capable of its own evolution is precisely an expression of the living God who is love, mature divine love.

In establishing the presence of the Creator Spirit throughout the natural world, the previous chapter concluded that far from being distant from the divine, the world is the dwelling place of God. This chapter works in a similar manner, charting a path from the activity of that same Spirit who is love to the correlative insight that the evolving world, operating without compulsion according to its own dynamics, works freely with the incomprehensible God in bringing forth the fullness of its own creation. Recall how the sequence of biological evolution occurs over billions of years. Life moves in the direction of complexity from single-cell creatures to whole populations of plants and animals interrelated in dynamic ecosystems. Each increase in complexity is evoked by some mutation that successfully builds on a bodily structure already present and fits organisms to survive and reproduce better in a given environment. The advent of these mutations does not follow any neat time-table or logical prescription; they are genuinely random. In this aspect of its story the living world shares in the flexibility of the cosmos as a whole, which knows a certain indeterminacy all the way down to the sub-atomic level. While unpredictable in advance, however, the sequence of life's development can be reasonably understood in retrospect as the working out of innate propensities with which the universe is gifted from the beginning.

In a theological perspective, this whole process is empowered by the Creator who as love freely gifts the natural world with creative agency. Its relation to the living God is marked simultaneously by ontological dependence and operational autonomy. There is no rivalry. How could there be when the Creator Spirit is not a categorical being among other beings but the vivifying Source of all that exists. In, with, and under nature's own processes, God continuously creates the world. Correlatively, the natural world freely partners the Giver of life in the work of creation, moving through time with its own integrity.

THE WISDOM OF PHILOSOPHY: ULTIMATE AND PROXIMATE CAUSES

In the contemporary dialogue with science, a large cadre of theologians endorse the idea of nature's independent working as shown by evolution. They diverge mightily, however, over how to think about the relation between divine and created agency in an evolutionary world. The language of faith holds that one God created the world and operates within it in a personal, providential manner. At the same time, the natural world evolves according to its own intrinsic processes. How to think of the faith confession without compromising the integrity of what science has discovered? How to acknowledge genuine agency on the part of creatures without setting up competition with their Creator? We must be clear that the way God acts escapes total rational analysis since it shares in the ineffable mystery of God's own being. Nevertheless, theology jumps into the breach here, seeking some rational way to mediate comprehension .

A brief review of some major theories will reveal the fecundity of thought going on in this area. The thumbnail sketches below are far from doing justice to the complex nuances of each position in itself; some scholars combine more than one position in their own proposals. While representing the mere tip of the iceberg in each case, these descriptions indicate ways of thinking about God's way of acting in the world that will set the stage for discussing one more option. This will be the classical notion of primary and secondary causality, related to the dynamic of participation in being, a position which I find to be rich with interpretive possibilities.[7]

❧ Holding that God works in the whole of cosmic history and not just its initial design, and thus differing from deism, *single action theory* discerns God as the agent whose intention is carried out in the overall development of the cosmos, rather than in its particulars. The whole evolutionary development can be interpreted as "one all-encompassing action," unified by God's intention. This position draws from philosophy of action's elaboration of the relation of agents to their actions, through which they realize their intentions (Gordon Kaufman, Schubert Ogden, Maurice Wiles).

❧ Drawing on information theory, the *top-down causality* theory understands that God acts by feeding a flow of information into the system of the world-as-a-whole, influencing its operation the way a patterned whole influences the parts that belong to it. Whole-part influence in natural systems explains, for example, why a carbon atom acts so differently in a diamond and in a green plant: the whole in which it exists influences the part. By analogy, divine top-down input into the world's system affects the world at large without abrogating any of the laws of nature. This position finds an affinity with the Johannine concept of the *logos*, the divine Word, which may be taken to emphasize God's creative patterning of the world from the beginning and, in Christ, also God's self-expression *in* the world (Arthur Peacocke).

❧ The *'causal joint' theory* uses the innate openness of physical processes to predicate that God inserts divine influence at significant hinge points in open systems to actualize one of the many possibilities present. By introducing a specific determination at the quantum level, or into non-linear dynamic systems, or even into genetic openness, divine act influences the overall outcome; by deciding these indeterminacies, God makes the divine will effective in the world (Nancey Murphy, Robert Russell, George Ellis, John Polkinghorne).

❧ Critical of mind/body dualisms, the *organic* model envisions the world as the body of God. On analogy with the agency of embodied human persons, this position envisions the spirit of God acting universally and particularly in the world the way the personal self acts in and through one's body. Divine presence acting with transcendent immanence in the body of the world emphasizes deep connection rather than the distance of the monarchical model. "God's action as the spirit of the body is twofold. The spirit is the source of life, the breath of creation; at the same time, the Holy Spirit is the source of the renewal of life, the direction or purpose for all the bodies of the world – a goal characterized by inclusive love."[8] (Sallie McFague, Grace Jantzen)

❧ The *kenotic* position perceives that God voluntarily self-limits divine power in order to participate vulnerably in the life of the world, making room for its freedom the way a parent's patient, self-emptying love enables a child to grow (John Hick, Keith Ward, Paul Fiddes, John Haught).

❧ Laying out a coherent metaphysical system, the di-polar theism of *process thought* sees God as a creative participant in the cosmic community who acts in all events by influence or persuasion. In providing initial aims to every "actual occasion" or developing event, God lures the world in a desired direction toward new possibilities of a richer life together (Charles Hartshorne, John B. Cobb, Ian Barbour, David Griffin).

While markedly different from each other, these various positions have much in common. They share a profound respect for the freedom of the natural world to evolve consistently with its internal laws as discovered by contemporary science. They eschew positing a God of the gaps, brought in to explain what science has not yet figured out. Just as strongly they shun an interventionist view of divine activity which posits God's influence on natural events in a way extrinsic to their own inner working. All are proposing models in which the creating God known in an historic faith tradition might be understood as acting in nature known by evolutionary science. In different ways they seek to make intelligible the idea that the creating God as ground, sustaining power, and goal of the evolving world acts by empowering the process from within.

It may seem that the theory of primary-secondary causality joins this discussion etched with the appearance of hoary old age. Elucidated by Aquinas and other scholastics, its origin in a static worldview would seem to disqualify it from interpreting creative agency in a natural world that evolves. Yet the opposite proves to be the case, for the basic principle remains fruitfully the same: the creative activity of God is accomplished in and through the free working of secondary causes. Science may describe these causes today in ways that differ from the static cosmos of medieval thought, but they can still be interpreted as the means by which God fulfills divine creative purpose.

The limits of language frustrate what we are trying to say here. The God-world relation is unique, and the primary-secondary causal dynamic is applicable only in this instance. Thus we must be clear that these two causes, ultimate and proximate, are not two species of the same genus, not two different types of causes united on a common ground of generating effects. They operate on completely different levels (itself an inadequate analogy), one being the wellspring of Being itself, the Cause of all causes, and the other participating in the power to act, as things that are burning participate in the power of fire. The relation precludes competition precisely because the living God, "source and goal of all things,"[9] is not included among the "all things" that work by natural laws. The horizon cannot be included within the horizon. David Burrell makes the point pithily: "the creator in acting acts always as creator."[10] And how is that? By continuously gifting creatures with their existence and power to act.

When Aquinas uses this philosophical construal to explain how God's creative purpose is achieved in the course of time, it enables him to posit a strong notion of the natural world's autonomy. He is so convinced of the transcendent mystery of God (pure aliveness, *esse*) and so clear about the unique way God continuously creates the world, that he sees no threat to divinity in allowing creatures the fullest measure of agency according to their own nature. It is characteristic of the creative power of God to raise up creatures who participate in divine being to such a degree that they are also creative and sustaining in their own right. A view to the contrary would diminish not only creatures but also their Creator. In Aquinas' words, "to detract from the perfection of creatures is to detract from the perfection of divine power."[11] It belongs to the perfection of creatures to act according to the fullness of their abilities, as finite causes. This gift the courteous Creator bestows without reserve: "the dignity of causality is imparted even to creatures."[12]

As is the case with creatures' participation in divine being and goodness, so too with agency. The power of creaturely agents to cause change in the world is a created participation in the uncreated power of the One who is pure act. Conversely, the Creator Spirit's generous goodness and wisdom are seen especially in the creation of a world with its own innate agency. This is not to say that God's action becomes part of the creature's essential action, which has its own integrity. On the contrary, God's act giving creatures their very nature is what makes autonomous creaturely action possible at all. The Source creates and sustains, while creatures receive their form and power to act with their own efficacy.

Such a way of thinking does not require that God as ultimate cause work in the world apart from secondary causes, or in addition to them, though its logic does not prevent this.[13] In terms of evolution, this view does not envision that God's divine act supplies something that is missing from a creaturely act, or secretly replaces it so that creatures are only a sham cause. Nor does it see that divine and finite agents are complementary, each contributing distinct elements to the one outcome. In truth, God's act is not a discrete ingredient that can be isolated and identified as a finite constituent of the world. In this sense the world necessarily hides divine action from us. The living God acts by divine power in and through the acts of finite agents which have genuine causal efficacy in their own right. The wonderful word *concursus*, meaning flowing or running together, comes into play to express this idea. Far from being merely a tool, instrument, or puppet in divine hands, the world acts with its own free integrity to shape its own becoming. It is empowered to do so by the transcendent mystery of the Spirit of God, who pervades the world, quickening it to life and acting in and through its finite agency.

In Aquinas's discussion of divine governance of the world, both divine primary causality which encompasses the world and creaturely secondary causality which participates are correlated with final causation, a creature's innate tendency toward a goal, to provide a grid for understanding. It would seem, he objects with a curiously modern ring, that the universe does not need to be governed by God, for the processes of the world seem to accomplish their purpose on their own and without any interference. However, this very self-direction is itself an imprint (*impressio*) from God, for in giving creatures their own being God gives them a natural inclination whereby through their own actions they tend toward a goal. This dynamic tendency is genuinely part of their own nature which at the same time expresses God's purpose. While endowing creatures with their inbuilt nature and ways of acting, God leaves them free to follow the strivings of their natural inclination which aims them toward a natural good. Since all good is a participation in divine goodness, the universe as a whole tends toward the ultimate good which is God. In scholastic categories this is summed up in the notion that God is immanent in the universe as final cause. Pleasingly, Aquinas finds that this view resonates with the biblical depiction of Holy Wisdom, who reaches from one end of the world to the other, ordering all things sweetly and mightily (Wis. 8.1).[14]

Let us draw these threads together to see how they might deliver an interpretive view of evolution's autonomous workings while affirming the Creator Spirit's innermost presence and action. As Aquinas explains, the way God is governor of things matches the way God is their cause. God is ultimate cause of the world as a whole and in every detail, endowing all created beings with their own participation in divine being (enabling them to exist), in divine agency (empowering them to act), and in divine goodness (drawing them toward their goal). Immanent Ground of all, God's intention comes to fruition by means of purposes acted out in those who are thus grounded. Why is this fitting? Aquinas argues in a particularly insightful reply that those forms of governing are best that communicate a higher perfection to the governed. Now there is more excellence in a thing's being a cause in relation to others than in its not being a cause. Consequently, God governs in such a way as to empower creatures to be causes toward others. Indeed, "If God were to govern alone, the capacity to be causes would be missing from creatures,"[15] to the detriment of their flourishing and their Creator's glory. Looked at another way, if God did everything directly so that created causes did not really affect anything, this would be a less powerful God. For it shows more power to give others a causative capability than to do everything oneself.[16] The great-hearted God imparts to creatures the dignity of causing.

It seems to me that it is so easy to forget this, slipping God into the web of interactions as though the divine were simply a bigger and better secondary cause. But the philosophical distinction between ultimate and proximate causality enables thought to hold firm to the mystery of the greatness of God and the integrity of creatures in equal measure. Everywhere present and active, the Creator is not an individual factor among others that bring forth species. Instead, the Spirit of God continuously interacts with the world to implement divine purpose by granting creatures and created systems their full measure of efficacy. This is a both/and sensibility that guarantees the integrity of the created causal nexus while affirming the gracious and intentional immanence of the transcendent God active within worldly purposiveness. To my way of thinking, it is a technical way of interpreting how mature Love acts.

In the dialogue among various contemporary positions on divine action, the primary-secondary causality position, usually called neo-Thomism, receives criticism on several fronts. Assessing its strengths and weaknesses, Ian Barbour notes that while it has the great merit of respecting the integrity of the natural causal nexus, it has difficulty in moving away from divine determinism to allow for genuinely random acts to occur;[17] however, if chance be given the status of a secondary cause, this problem disappears, as will be discussed below. In a related objection, Arthur Peacocke notes that proponents of this position have on occasion used the artisan/instrument analogy to explain how God the primary cause works through secondary causes in the world.[18] The lumberjack uses an ax to chop down a tree: the active agent uses a tool, both accomplishing the goal in different ways. The problem with this analogy, Peacocke rightly observes, is that it completely overlooks the independent operation of natural causes in the world which act with their own inner dynamism; nature is not an ax, an inert tool. A more fundamental problem with the artisan/instrument analogy is that it conflates what is technically called an instrumental cause (the ax) with secondary causality. When the subject is creation, however, both the lumberjack and the ax are secondary causes. The Creator does not relate to the world the way a carpenter uses her hammer or a sewer his needle. Rather, the Spirit of God sets the world up in the fullness of its own efficacious powers which are grounded in the gift of being created.

Other critics argue that this theory merely *asserts* that God acts through natural causes without giving any idea of the mechanism by which God's purpose is accomplished. Without the "how," it is said, this is a position without any explanatory power. One of the chief levelers of this criticism, John Polkinghorne, argues that primary-secondary causality actually offers nothing at all that illuminates how God acts in the world.[19] His own proposal for a causal joint connection, as noted above, argues ingeniously that thanks to the indeterminism of reality at many levels, God's direct intervention in any instance does not transgress the laws of nature. Natural systems themselves are "gappy" and open enough to receive outside influence without being violated. God could manipulate indeterminate quantum events, for example, deciding the instant at which a particular radioactive atom decays; or in answer to prayer God could arrange for the sun to shine on the church picnic by setting certain initial conditions in the weather pattern a week ahead. By acting within micro-events, subsequently amplified, divine action could thus affect outcomes without having to set aside natural laws.

This position, it seems to me, commits a double fallacy. On the one hand, regarding nature, the openness of dynamic systems to a variety of outcomes is an intrinsic structure necessary for the integrity of their own operation. The soundness of their function requires that their indeterminacies be decided by natural means. In principle there are no gaps in the universe that the transcendent God can quietly slip in to fill. Inserting hidden divine action into an open system thus compromises the natural order. In principle this is no different from the classical idea of divine intervention in a rigidly law-controlled world, except that such intervention is now hidden. On the other hand, regarding the incomprehensible mystery whom we call God, this causal joint position errs by making God into a bigger and better secondary cause. But the living God is not part of the causal nexus of the created world. Inserting divine action into indeterminate systems reduces holy Mystery who creates and sustains the whole world to a bit player. To the contrary, as Edward Schillebeeckx astutely observes, "Belief in God the creator is never an explanation, nor is it meant to be."[20] It is good news, liberating good news about the gracious presence of God in and with the finite. The Creator brings autonomous, non-godly beings into existence, abides with them, and wills to be their God, even in their finitude, which is not a flaw. Belief in the Creator God delineates the ultimate meaning of the universe, not an explanation of how things work.

My own assessment of the philosophical meaningfulness of the ultimate/proximate causality dynamic holds that it reasonably refutes criticism only if positioned within the overarching notion of the Creator God as the absolute Living One, pure wellspring of being, and the concomitant notion of creaturely participation. Taken cold without these roots it does not function with high wattage clarity. With these foundational pieces in place, it fairly resonates with potential to account for the full play of natural causes. Scholars who work with the neo-Thomist position consistently register how it lets the world be the world and evolve in its own way. "Aquinas, of course, had no notion of the evolution of species," writes Herbert McCabe; but seeing this process as a typical manifestation of the wisdom of the Creator, "he would I am sure have been delighted by the sheer simplicity and beauty of the idea."[21] Denis Edwards concurs: "Thomas Aquinas long ago clarified that God's way of acting in the world (what can be called primary causality) is not opposed to the whole network of cause and effect in nature (secondary causality). God's work is achieved in and through creaturely cause and effect. It is not in competition with it. Aquinas never knew Darwin's theory of evolution, but he would have had no difficulty in understanding it as the way that God creates."[22]

There is yet more to be said, but this foray into current philosophical discussion can provide thought structures that make room for the entangled bank to exist with the freedom of its own ways within the vision of faith. Two agencies of infinitely qualitatively different magnitudes are present in the same worldly action: the autonomous creaturely agency which enacts it, and the divine agency which founds, sustains, and empowers it. These are not two actions doing essentially the same thing, acting in a parallel

way, each contributing to part of the effect. Brought to life by divine generosity, creatures are genuine centers of activity that operate with their own
causal efficacy, interrelated and dependent on each other as well as on God,
while the ineffable, transcendent Mystery dwelling within the evolving
world continuously creates through the world's own autonomous processes,
letting it be and self-spending in an outpouring of love.

INTERPLAY OF LAW AND CHANCE

Bringing the primary-secondary causality construal to bear on the
evolutionary account of species makes possible a robust theological interpretation of what initially may seem one of its problematic aspects, namely,
random occurrences. The core of Darwin's theory speaks of variations
that occur spontaneously, he knew not how, and of natural selection that
continuously affirms or rejects them. Variations, what today we would call
genetic mutations, are an example of chance events. Natural selection is
an instance of an orderly principle which operates with regularity, akin
to a law. Bridging two language games, we can identify chance and law as
chief secondary causes at work in the evolution of life. "It is the interplay
of chance and law which is in fact creative within time," Peacocke writes,
"for it is the combination of the two which allows new forms to emerge
and evolve."[23] A closer look at this dynamic puts the theological validity of
chance into bold relief.

   Law refers to a orderly suite of natural forces that govern how the
universe works. These principles, read off from the regularities observed in
the world, hold true in all ordinary circumstances. Drop an apple, it falls
to the ground due to the law of gravity. These inherent regularities of the
world go back to the basic contingency that certain constants, processes,
and relationships emerged as the universe developed over time. William
Stoeger makes the important point that in the emergent universe, laws of
nature should be understood as descriptive rather than prescriptive, that is,
they are descriptions read off from regularities in the universe that approx-
imate what we observe, rather than rules that preexist platonically prior to
and apart from the universe which operate to dictate behavior.[24] Laws of
nature guarantee a steady dependability in view of which we can calculate
and make predictions. As Einstein famously put it, "the eternal mystery of

the world is its comprehensibility," seen in the reliable laws of physics and chemistry spelled out in mathematical formulas.[25]

Chance refers to the crossing of two independent causal chains that intersect for no known reason that can be figured out in advance. These random events interrupt the necessary regularity that lawful systems establish. The interruption may be destructive or it may open the possibility that something new might emerge from within these systems. Either way, things do not go on as before. Ancient stars explode in a cataclysm that brings about a new solar system; an asteroid hits the Earth and wipes out the dinosaurs; a gale wind blows some birds off course to an uninhabited island: the unpredictable, uncontrollable character of chance makes the history of life shot through with surprise. One might riff on Einstein and say that along with its comprehensibility, what is equally mysterious about the universe is its unexpected open-endedness.

Together: In the evolutionary process, changes in genetic material bring about changes in the structures of organisms, which in turn make possible new behaviors and relationships. These mutations are inherently unpredictable at the molecular level at which they occur, and are random with respect to the needs of the organism: many are harmful, some few beneficial. Uncertainty also awaits in the particular environmental niche where the mutated organism has to interact in the struggle for food, mates, and the avoidance of predators. Chance appears in both the internal mutation and the external environment. Far from creating a confused jumble, however, these random events operate within a milieu which constrains and delimits their possible outcomes. Natural selection screens out mutations maladapted to the environment and preserves those that are beneficial. It does so in ways that are amenable to regular statistical description in populations as a whole.[26] Without such constraints small changes would dissipate in chaos. With such selection in place, random changes are accommodated in ways that allow regular trends to take root and develop. Bit by bit over tens of thousands of generations successful functions are selected, weaker genotypes are weeded out. The biological world becomes ever more beautiful and complex. The dance of law and chance over unimaginable eons of time brings forth the community of life on Earth as we know it.

If all were law, the natural world would ossify; its ordered structure would be rigid, repetitive, deterministic. If all were chance, nature would dissolve in chaos; no new patterns would persist long enough to have an identity. But chance operating within a lawlike framework introduces novelty within a pattern that contains and directs it. Their creative interplay brings forth ever-new living forms. Rather than being an enemy of law, then, chance is the very means by which nature becomes continuously creative.

Compared to law, however, chance poses a challenge to thought precisely because it is so unpredictable. There are important philosophers and scientists so struck by the iffy occurrence of chance and its uncertain outcomes that they have elevated the play of chance to a metaphysical principle. The fact that any creature comes into being at all, the argument goes, is the result of purest accident. Consequently, any idea that the universe has an overall direction or purpose must be false, along with the belief that there is a Creator God engaged with the process.[27] In response, other thinkers call attention to the fact that chance is not the only dynamism at work in evolution. Random events occur within organisms, populations, and ecosystems that display regular trends over time. In this camp, Peacocke opens a way ahead with a striking idea: why not see chance as a tool that allows matter to explore the full range of its possibilities:

> Instead of being daunted by the role of chance in genetic mutations as being the manifestation of irrationality in the universe, it would be more consistent with the observations to assert that the full gamut of the potentialities of living matter could be explored only through the agency of the rapid and frequent randomization which is possible at the molecular level of the DNA.[28]

The capacity to form a living world is there from the beginning in the fundamental constitution of matter-energy and its emergent laws. Chance mutations are the way the stuff of the universe gets investigated, its potential unpacked, so that it moves in the direction of living richness and complexity. The fact that life as we know it is inseparable from the unforeseen events that mark its history simply places our planet squarely within the dynamism of the wider universe.

To digress to the human species for a moment, it is a given among philosophers of science that the emergence of human nature is based on the existence of a natural infrastructure of this kind. There is a deep compatibility between the creative though not conscious ways physical, chemical, and biological systems operate though the interplay of law and chance on the one hand, and persons' experience of consciousness and freedom within their physical embodiment on the other. These particular human qualities are intensely concentrated states of capacities found throughout the universe in natural forms. At the very least, the freedom of natural systems to explore and discover themselves within a context of lawlike regularity is one of the natural conditions for the possibility of the emergence of free and conscious human beings as part of the evolving universe.

What sense can theology make of this dynamic so basic to the biological world? At the outset it can simply be said that regularities and their chance interruptions are secondary causes. Through their reciprocal operation the Spirit's creative purposes are being realized. Propensities given to creation by the Creator in the beginning are gradually actualized by the operation of chance working within lawlike regularities over deep time. "God is always acting through the deterministic and indeterministic interrelationships and regularities of physical reality," Stoeger writes, "which our models and laws imperfectly describe."[29] Peacocke goes a step further in observing that the interplay of law and chance are what one would *expect* if to begin with the universe were sent rolling through time with the power to figure out its own way. With this insight we have a vocabulary for expressing the creative activity of the Spirit and the natural world's freedom in its own making.

If law stands for the constants of the world, for its steady physical properties and regular processes, then this regularity can be regarded as a feature with which God has endowed the world. If chance stands for the unpredictable interruption of this regularity by other natural forces, then this capacity for surprise can also be taken as a God-given feature of the world. From the beginning the possibility of becoming "more" is written into creation. The operation of chance explores possibilities, stimulating new forms to come into existence. The interaction of chance and law becomes a creative means, over time, for testing out, tweaking, and finally evolving every new structure and organism of which the physical cosmos is capable. It is, as Peacocke astutely observes, what one might *expect* if God created the world to be a participant in developing its own richness.

In turn, the interplay of chance and law allows us to infer indirectly something about the One who creates the world with this dynamism. Theology has traditionally allied God with lawful regularity, seeing in the reliable, intelligible features of nature an expression of divine will and purpose. This is still a fine idea. The deep regularities of the world in their own finite way reflect the faithfulness of the living God, reliable and solid as a rock. It has been more difficult for chance to find a home in the theological imagination, especially if this be governed by the model of the monarch. Given evolution's capacity to surprise, however, theology can now make a capacious affirmation. The occurrence of chance in the world in its own finite way reflects the infinite creativity of the living God, endless source of fresh possibilities. The indwelling Creator Spirit grounds not only life's regularities but also the novel occurrences that open up the *status quo*, igniting what is unexpected, interruptive, genuinely uncontrolled, and unimaginably possible. As boundless love at work in the universe, the Spirit embraces the chanciness of random mutations, being the source not only of order but also of the unexpected breaks in order that ensure freshness. Divine creativity is much more closely allied to the outbreak of novelty than our older order-oriented theology ever imagined. In the emergent evolutionary universe, we should not be surprised to find the Creator Spirit hovering very close to turbulence.[30]

UNSCRIPTED ADVENTURE

The interplay of law and chance over deep time underscores the fact that the history of evolution is amazingly unscripted. The origin of species does not necessarily follow any neat logical plan but is shot through with surprise. A favored imaginative game among scholars in the field is to rewind the tape of life's evolution back to the beginning, and let it roll again. Would the community of life look as it does now? No. Millions of small biological events would never repeat in the exact same way at the same time, and while the eye and the wing might well emerge again since they have done so many times in the course of evolution, the precise figuration of bodies and relationships of ecosystems would be different.[31] Seen retrospectively, an intelligible story of life's emergence can be constructed, which is what Darwin did. But prospectively there is no telling what might happen. Thus the overall arc of life on Earth is not properly described by analogy with an acorn growing into an oak tree or an embryo developing in the womb, with all aspects of the mature creature inscribed in advance. A better analogy might be a wild ride through time whose outcome defies prediction. Far from being a pre-programmed machine, the biological world tends toward richness and diversity through the outworking of its own creative self-organization. Holmes Rolston's wry observation is apt: the laws of physics and chemistry are reliable, but nothing in them demands that Earth be created, let alone with elephants.[32] It is all such an adventure.

The creative agency of the Spirit of God does not shut down this openness, but enables it. God does not act like a bigger and better secondary cause determining chance atomic events, or initial conditions of chaotic systems, or genetic mutations. Rather, divine Love empowers the structure of creation which operates with its own integrity, all the while supporting unfolding events as they weave into regular patterns toward the realization of an ever more complex whole. Jesuit astrophysicist George Coyne puts it simply: God in "infinite freedom continuously creates a world which reflects that freedom."[33] God lets the world be what it will be, he goes on, not intervening arbitrarily in its evolution but participating, lovingly, in its becoming . Creative divine sovereignty and creaturely freedom, of which chance is one instance, do not compete. To the contrary: the genuine interplay of chance and law in the unscripted evolution of life is due to the generous way the Giver of life creates the world. Thanks to this gracious Love, the natural world freely participates in its own creation.

EMERGENCE: ON BEHALF OF MATTER AND THE BODY

This theological interpretation of how evolution grows the tree of life calls for a new understanding of matter. Classical ideas that cast matter and its concrete bodily forms into categories such as inert, "dead," or inferior to spirit no longer hold up. As described by the natural sciences, matter has evolved from inanimate to animate to intentional states. Over eons of time levels of complexity mount from atoms to molecules (think of how two atoms of hydrogen coupled with one of oxygen produce water, a very different substance), and thence to genes, cells, tissues, organisms, species, ecological populations. The "more" that appears in each case is not something added on externally to what was there before but something that emerges from within as a result of nature's interactions at every level. In an ordinary sense the term "emergence" connotes something coming out of hiding, coming into view for the first time. Evolutionary scientists use it to describe the spontaneous appearance of unprecedented new biological forms. In the emergent universe, gravitational attraction of dust and gasses has produced planets; molecular and chemical interactions have led to living cells; the process of natural selection has acted on living systems to create organisms with high-functioning consciousness. In each case the emergent phenomenon gathers up what has preceded it, shaping this material into a new, more complex unity. What emerges has distinctly different properties and functions from what went before, though still composed of the same fundamental matter. Matter has become patterned in a new way. Organized with more complexity, it develops new properties, acts with novel powers, enters into more diverse networks of relations. As a result, new levels of reality appear over time that require new language and concepts capable of describing them accurately; the laws of physics and chemistry alone do not completely describe biological phenomena. Emergence signifies that more comes from less due to the fecundity of nature itself.[34]

Pondering this reality, Karl Rahner proposes that we embrace a fundamental idea: matter has the capacity to transcend itself. Matter can do this because it has been endowed by its Creator with an inner tendency, a quiet, powerfully pulsing drive, to become something more. The foil against which Rahner places this dynamic assessment of matter is the philosophical dualism which radically separates matter and spirit, considering matter passive and devoid of movement. If we take our cue from how spirit and matter are related in a human person, however, as Rahner does in his systematic anthropology, a different understanding ensues. A human person is a dynamic unity of matter and spirit, an embodied spirit in the world. Far from the body being a dispensable container for the soul, corporal and spiritual dimensions form one unified being. Humans experience themselves as a unity in the way they know and question, with their physical senses interacting with their mind, and the way they desire and love which likewise engages bodily and spiritual dimensions. Furthermore, as physical beings, persons are able to go beyond themselves toward infinite mystery in every intellectual question, every act of love. Rahner allows the unified experience of the human subject to cast its light back over the natural world from which have humans emerged: "Starting from this inner interrelation between these two factors [matter and spirit in the human being] and concentrating on the *temporal* duration of this relationship between these two factors, it may be said without scruple that matter develops out of its inner being in the direction of spirit."[35]

Rahner's concept of matter's active self-transcendence correlates with the notion of biological evolution as an ongoing process where something new indeed emerges. Darwin's completely non-transcendental description puts this dynamism in concrete terms when, reading old treatises on hyacinths, potatoes, dahlias, etc., he observes, "It is really surprising to note the endless points in structure and constitution in which the varieties and sub-varieties differ slightly from each other. *The whole organisation seems to have become plastic*, and tends to depart on some small degree from that of the parental type" (12, italics mine). Far from being an inert substance, matter has a dynamic urge to explore; it is oriented to become more. Pressing forward with a power interior to itself, it can move beyond itself in the present moment toward ever more complex forms, even crossing thresholds to become new natures never before seen. Do not underestimate what it means for nature to become, urges Rahner. True becoming entails that nature surpasses itself, attains a greater fullness of being, reaches an inner increase of being proper to itself ... and does this not by adding something on but from within. God made the world this way, conferring on creatures an extraordinary capacity for becoming more. In theological language, this is called:

> God's conservation of the creature and ... concurrence with its activity, in the inner and permanent need of all finite reality to be held in being and operation, in the being of becoming, in the being of self-becoming – in short, in the being of self-transcendence which belongs to every finite being.[36]

Sharing the concern of the Thomistic tradition to preserve the integrity of secondary causes, Rahner stresses that in evolution the activity of divine presence must be thought of as something so *interior* to the creature "that the finite being is empowered by it to achieve a really *active* self-transcendence and does not merely receive this new reality passively as something effected by God."[37] It is not the case that the Creator acts as a categorical cause which intervenes at certain points in the time and space of evolutionary history without any inner-worldly causality being operative at all. Far from being a finite cause side by side with others in the world, God is "the living, permanent, transcendent ground of the self-movement of the world itself."[38] Hence, "under the dynamism of divine being and under the continuous divine creative power,"[39] the material stuff of nature evolves in the integrity of its own processes. If this be granted, then even "the self-transcendence of the inorganic into life (always supported by the creative but transcendent, uncategorized dynamism of the absolute being of God) may be seen and acknowledged."[40] Matter can transcend itself at any moment in bringing forth life and ever new forms of species. This idea, Rahner suggests, is the secret of life. It offers one more intellectual explanation that supports belief in God as Creator while appreciating science's finding that matter has evolved in the direction of life and then consciousness under its own steam. Nature's capacity for active self-transcendence is the key. In the realm of biological evolution, natural selection's work on spontaneously arising variations gets that key turning to throw open the door to ever-new forms of life. Novelty comes about by the self-organizing dynamism inherent in creatures themselves. Evolution over deep time is so creative because the material of the world itself has the God-given inner ability to become ever more.

BEASTS AND ENTANGLED BANK

In the course of thinking upon these things, theologians are finding it helpful to imagine new metaphors to capture the nuances of the Giver of life's creative relation to the autonomous workings of the cosmos. As might be suspected, these images are drawn more from artistic experience than from the classical models of a monarch giving commands or an artisan plying inert tools. No one of these metaphors, of course, is adequate but each sheds a little light. Among them: the Creator Spirit is like a composer of a fugue, who starts with a simple line of melody and then weaves a complex musical structure by endlessly folding it back upon itself; or like a jazz player, inspired by the spirit of the audience and the night to improvise riffs upon a basic melody; or like a theatrical improvisor of unsurpassed ingenuity in live performance, who amplifies and embroiders each theme as it presents itself; or like a choreographer, composing dance steps in tandem with the creative suggestions of the whole troupe; or like a game designer who salts the deck with wild cards. In each of these examples the image is arrived at through the logic set out in the philosopher W. Norris Clarke's evocative passage:

> What must the "personality" or "character" be like of a Creator in whose image this astounding universe of ours is made, with its prodigal abundance of energy, its mind-boggling complexity, yet simplicity, its fecundity of creative spontaneity, its ever surprising fluid mixture of law and chance, etc. Must not the "personality" of such a Creator be one charged not only with unfathomable power and energy, but also with dazzling imaginative creativity?[41]

Each thinker who crafted one of these metaphors is trying to make room in the religious understanding of creation for the surprising way life has evolved by the inner workings of the natural world itself. The foil against which the metaphors work is the image of a heavenly ruler who exercises direct supernatural control over everything that happens on Earth, vitiating the integrity of natural causes. As long ago as the fifth century Augustine noted that even the Genesis story gives the natural world a role to play: God says "let the waters bring forth" (1.20) and "let the earth bring forth" (1.24), and they do. The theory of evolution today rachets up what the sea and the earth can do. The theological challenge is to seek an understanding of faith that renders fair account of the intense creative activity of both Creator and creation.

To my mind, a theology of the Spirit as the love of God in person indwelling the natural world and sparking its own daring generative powers goes a good distance toward meeting this challenge. Infinite mystery of self-giving love, the Creator Spirit calls the world into being, gifts it with dynamism, and accompanies it through the by-ways of evolution, all the while attracting it forward toward a multitude of "endless forms most beautiful" (490). We glimpse here bounteous personal love that pours itself out in empowerment of a creation that is transient and vulnerable yet resilient and generative, a creation that without this love would be literally nothing at all. As such unbounded love will do, the Spirit of God unleashes autonomy in the beloved rather than seeking to control the other by any form of power-over, even if benevolently exercised. Sheer overflowing goodness, the Creator respects the freedom and independence of the world such divine bountifulness lets loose, and works through its dynamisms and interlocking evolutionary processes.

In view of the openness of the natural world, the biblical images of the Spirit return with ever more significance. Blowing wind stirring up the world; flowing water saturating it with the juice of life; burning fire igniting its steady and unexpected events; brooding bird bringing it to life with the love of a mother's warm body; holy Wisdom, more mobile than any motion, pervading it with her ordering and renewing spirit: all evoke the insight that the creative Spirit of God desires free partnership, not subservience. Neither overriding monarch nor absent deist god, the Spirit of God moves with extravagant divine generosity to create and sustain the conditions that have enabled the biodiverse community of life to become so interesting and beautiful. The unimaginable epochs of time over which this has occurred are themselves a gift of opportunity for nature's emergent freedom to work.[42]

If we "ask the beasts," counsels the book of Job, they will teach us "has not the hand of the Lord done this?" (Job 12.9). If we "contemplate an entangled bank," as Darwin advises, it will become clear that its elaborate forms "have all been produced by laws acting around us" (489). At this point the testimony of both beasts and bank can meet on common ground. Far from being in competition with the laws acting around us, including natural selection, the hand of the God of love empowers the cosmos as it evolves these very laws and their emergent effects. The world develops in an economy of divine superabundance, gifted with its own freedom in and through which the Creator Spirit's gracious purpose is accomplished.

By such pathways of thought, a pneumatological interpretation of continuous creation, drawn from biblical and theological tradition, is one way to respect the discoveries of evolutionary theory while showing that belief in the God who creates is still seriously imaginable. Just as important, such theological reflection highlights the insight that the Creator loves the rich diversity of the tree of life, embedded in the whole tapestry of the cosmos, for its own sake, and not only as a stage on the way to the human species. To say this is not to detract from the singularity and importance of human beings. But it is to give value to the existence of the natural world, long neglected as a site of theological interest. Rather than devoid of Spirit, it is the dwelling place of God. Rather than a mere backdrop, it is the locus of divine activity deeply involved in, with, and under its open-ended evolutionary processes. Dante once wrote eloquently of "the Love that moves the sun and the other stars."[43] We can now continue that this Love also moves the origin of species, nearer to creatures than they are to themselves, acting immanently throughout the matrix of the freely evolving community of life.

7

ALL CREATION GROANING

> We know that all our mothers bear us for pain and for death. O,
> what is that. But our true Mother Jesus, he alone bears us for joy
> and for endless life, blessed may he be. So he carries us within
> him in love and travail, until the full time when he wanted to
> suffer the sharpest thorns and cruel pains ... and at the last he
> died. And when he had finished, and had borne us so for bliss,
> still all this could not satisfy his wonderful love.
>
> Julian of Norwich

"WE SUFFER AND DIE"

There is yet more to think about from a theological perspective,
for life evolves at a terrible cost in pain and death. The natural
world of living organisms is not just the beautiful dwelling place
of the Creator Spirit whose love empowers creation to evolve
according to its own free, rigorous processes. It is also a place of
agony insofar as these processes exact a high price. As Darwin tells
its history, the dynamic of evolution pushing toward ever more
complex and beautiful life forms entails struggle that brings pain
and suffering even unto death. The sheer extent is mind-boggling.
"Nature is random, contingent, blind, disastrous, wasteful, indif-
ferent, selfish, cruel, clumsy, ugly, full of suffering, and ultimately
death,"[1] as Rolston describes. Yes, but there has to be light to

cast shadows, and with evolution we are dealing with a blazing light, the fertile procreation of new life. Biologically, pain and death accompany the ongoing passage of life which is always "struggling through toward something higher."[2] The fact of the matter is that glorious life arises and is renewed in the midst of its perpetual perishing.

The apostle Paul did not know the theory of evolution; how could he. The unfinished state of the natural world, however, was obvious, and he wove this awareness into a passage about redemption that gives bedrock direction to the next step in our theological reflection. The present time is filled with sufferings, he reflects in Romans chapter 8, but these cannot be compared with the glory to come. Subjected to futility, "the whole creation has been groaning in labor pains until now" (Rom. 8.22), like a woman in childbirth straining toward the arrival of new life. We ourselves also groan, he continues, waiting for the redemption of our bodies (8.23). In the midst of this agony dwells the Spirit, interceding "with sighs too deep for words" (8.26). This is indeed a world shot through with pain and trouble. Paul has already established that hope springs afresh because the Spirit of God who raised Jesus from the dead dwells in us and will give life to our mortal bodies also (8.11). Now he weaves the natural world into the picture, writing that creation waits with eager longing for this moment because it too will share in the liberation: "the creation itself will be set free from its bondage to decay and will obtain the freedom of the glory of the children of God" (8.21). The good news is that no suffering or power, not even death, can separate us from the love of God brought near in Christ.

The above thin line of exposition barely does justice to the rich themes Paul develops in this text.[3] The bare highpoints, however, provide a sequence of ideas which allow for a theological interpretation of the cost of evolution. To wit: Creation is groaning. There is hope for deliverance. The love of God in Christ Jesus "who died, yes, who was raised" (8.34) grounds this hope. To round out the exploration of continuous creation this chapter links the Creator Spirit present and active in the world with the love of God made known in the death and resurrection of Christ, beginning with the groaning and then moving to the hope.

Pain enters the natural world with the emergence of neurons, nervous systems and brains. These specialized biological organs enable a creature to register and assess information coming in from its environment. The more precisely an organism can evaluate whether something is harmful, neutral, or helpful to its life, the better chance it will have of survival. Pain is a hurtful physiological stimulus that signals something is injurious; pleasure, a stimulus traveling along the same pathways, signs that something is beneficial. Both responses goad the organism to action. Pain encourages avoidance of what harms; pleasure induces engagement with what enhances well-being. Within rough limits both stimuli enable behavioral adjustment in the face of changing situations. Given the clear benefit for survival, sensitivity to pain and pleasure is selected for in the evolutionary development of life.

Suffering refers to an affective state of anxiety and anguish that arises in response to pain. While it is difficult to assess the mental states of animals with exactitude, contemporary veterinary studies are increasingly in agreement that suffering accompanies pain in organisms that have gained a certain level of awareness. As species evolve, nervous systems and brains grow more complex, allowing for heightened alertness, all the way to levels of consciousness typical of sentient animals. At this point physiological hurt triggers not only basic avoidance behavior but also emotional distress such as fear, anger, and grief stemming from the sense that something awful is happening. The more sensitivity in a species, the more suffering from pain and, concomitantly, the more joy from pleasure are experienced as consciousness ramifies throughout the evolving world. Capturing the newness of this phenomenon, Rolston describes suffering as "the shadow side of sentience, felt experience, consciousness, pleasure, intention, all the excitement of subjectivity waking up so inexplicably from mere objectivity."[4] The pathway to consciousness runs through flesh that can "feel" its way through the world. In that regard, suffering is irreplaceable.

Death, whether accompanied by pain and suffering or not, is a companion of life across the entire adventure. Every organism traces an arc through time which eventually comes to an end for itself as a single being, even if its species continues to exist. In their embodied vitality plants and animals play their role in the drama of life, interacting with the environment and other organisms. Then they die, vanishing forever as individual living entities, while the material of their bodies becomes nutrient for new life. On average life lasts a few hours for a mayfly, a few days for a daisy, ten years for a dog, hundreds of years for some trees, three score and ten for human beings, but however long the time span the biological life of the individual comes to an end either by accident, predation, or internal collapse. Like pain and suffering, death is indigenous to the evolutionary process. Without it, not only would there be no food for eaters to eat, but eventually there would be no room for new sorts of creatures to emerge. The time-limit that ticks away in all living organisms and ends with their death is deeply structured into the creative advance of life. Denis Edwards states the case simply: "Evolution demands a series of generations; ... without death there could be no wings, eyes, or brains,"[5] no soaring creatures in the sky, no fine-tuned eyesight, no advanced crafty brains.

Extinction of species rachets death up to an astronomical level. (I speak here only of extinctions that happen spontaneously in nature apart from the action of the human species; extinctions occurring today because of human actions present us with a murderous crisis that will be discussed later in this book). Recall *Origin*'s drawing of the tree of life and the layers of organisms that emerged, thrived, and disappeared in the time between the original parents and the descendant species fourteen thousand generations later. Reflecting on fossil bones of disappeared species he and others dug up around the world, Darwin compared the soil of the earth to a huge old museum that preserves and displays ancient curiosities. According to recent calculations, about 98 per cent of all previously existing species have gone extinct, some in global catastrophes of mass extinction. Those who live today walk upon this Earth as upon a vast cemetery. Therein lie the remains not only of individuals but of whole species of creatures. How stunning to think that massive death is intrinsic to the process of evolution. In the course of time it removes species less adapted to a changing environment, creating opportunities for more complex forms to emerge.

It is important to note that none of this agony and loss is due to human sin. Contemporary biblical scholarship enables us to read the third chapter of Genesis with its story of Adam and Eve, the garden and the serpent, in the same way we read the creation stories in Genesis 1 and 2, namely, as poetic, mythic narratives teaching religious truth about the relation of human beings and the world to God. Historically speaking, once life emerged there never was a literal garden of Eden or a paradise on this planet where death did not exist. Once nervous systems developed, there never was life without pain. Rather, as Peacocke observes:

pain, suffering, and death are present in biological evolution as a necessary condition for survival of the individual and transition to new forms long before the appearance of human beings on the scene. So the presence of pain, suffering, and death cannot be the result of any particular human actions, though undoubtedly human beings experience them with a heightened sensitivity and, more than any other creatures, inflict them on each other.[6]

And, we must add, inflict them on other species to a disgusting, sinful degree.

Considered in an evolutionary framework, pain, suffering, and death in the natural world do not fit into the common theological explanations offered for such occurrences among human beings. They are neither divine punishment for sin nor providential happenings intended to lead to soul-making or growth in virtue. Insofar as they are the result of the natural working out of life's creative processes, they are morally neutral. As John Thiel rightly observes, "nature lacks the personal character required for either guilty or innocent agency."[7] Orcas chase a sea lion through the waves, flipping it playfully in the air before devouring it; a lioness snags a wildebeest, knocking it down and biting its throat to cause asphyxiation; a hawk plummets to hook a scampering rodent with its sharp talons. The prey endures pain and death, but these are the result of interrelated life processes, not of some malign force. Indeed, benefits accrue. In every instance the nutrients in the lifestream of one organism are resources that nourish the life of the other. Over the long haul, the struggle to survive brings about rich, complex changes in structure and behavior. The sea lion species gains speed and agility as a result of the orcas' hunt; the weak wildebeest is culled from the herd and no longer reproduces; the rodents diversify their hunting habits. In a memorable image, "the cougar's fang has carved the limbs of the fleet-footed deer, and vice-versa."[8] Without pain, no further exploration of life's potential forms; without death, no new life. These afflictions arose as essential elements in a tremendously powerful process that created and continues to create the magnificent community of life on this planet.

And yet! The case of the backup pelican chick, increasingly used in theological discussion, brings the problematic aspect of evolution to a head in riveting terms. White pelicans ordinarily lay two eggs several days apart. The first chick to hatch eats, grows larger, becomes feisty. It tends to act aggressively toward the second-born, grabbing most of the food from the parents' pouch and often nudging the smaller bird out of the nest. There, ignored by its parents, the younger chick normally suffers starvation and dies despite its struggle to rejoin the family. Before this denouement, there is a window of opportunity in which, should some crisis befall the older chick, pelican parents can raise the second offspring and thereby have a successful reproductive season. It may also happen that in an especially good year the parents will feed and raise both chicks. But ordinarily the backup chick has only a ten percent chance of surviving. It is born as insurance. For the pelicans as a species this has been a successful evolutionary strategy, enabling their kind to survive for thirty million years. As depicted on video and shown on television, however, the ostracized chick's pinched face, small cries, desperate attempts to regain the nest, and collapse from weakness to become food for the gulls is a scene of such distress as to call for an account of this suffering in a created world considered good, the more so as the anguish of this one little creature is continuously repeated on a grand scale.

Could the biological world have developed otherwise? One might envision alternatives, though the majority of scientists, philosophers, and theologians working on this question hold that the correlation of pain and suffering with consciousness seems inevitable in a system where organisms interact with their environment. So, too, death and extinction are intrinsic to an evolutionary process that over thousands of millennia brings forth ever new forms, including human beings. Connecting this with Christian symbols Rolston describes the natural world as cruciform and its evolutionary process as a way of the cross, observing, "In general, the element of suffering and tragedy is always there, most evidently as seen from the perspective of the local self, but it is muted and transmuted in the systemic whole. Something is always dying, and something is always living on."[9] The laws known to us which have brought about the entangled bank, so pleasing in its beauty, have also rendered it a place of pain and death. Whatever might be true in other imaginable worlds, such adversity is in fact part of the story of life as it has evolved on our planet.

FRAMING THE ISSUE

Theological reflection on the natural world's continuous creation in the power of the Spirit cannot ignore this unfathomable history of biological suffering and death extending over hundreds of millions of years. Its overwhelming power initially evokes the honest response of being struck dumb in the face of so much agony and loss. As with the mystery of suffering among humans, its roots reach deeper than the human mind can fathom. When theology does dare to speak to this issue, ancient in its pedigree but relatively new in its evolutionary colorings, various viewpoints are endorsed and debated. The position that I am exploring in this chapter differs from two other streams of conversation in important ways.

First, this is not an exercise in theodicy. The theodicy project is a philo- *1)* sophical effort to construct a rational defense of God's goodness and power in a world where evil occurs. It figures there are reasons for suffering rooted in the divine will, and assesses pain, suffering, and death in such a way that they are reconciled with divine intent. The result is a somewhat satisfying intellectual system that justifies God by explaining reasons for suffering, making room for evils to exist in a logically meaningful world. A number of contemporary theologians are strongly critical of this theological effort, for compelling reasons. Theodicy, the criticisms charge, tends to gloss over the effects of evil in the concrete, denying the terror of innocent suffering and depriving victims of their voice; it ignores political practices and received social structures that cause enormous suffering, thus allowing the suffering to continue; by attributing to God what in fact is evil done by humans in the history of injustice, it induces passivity and cuts the moral nerve of protest. In a word, theodicy attempts to rationalize what is in fact a deep mystery beyond comprehension, with deleterious practical effects.[10] My own sense is that suffering and death are too much of an enigma to submit to such logic. Rather than a theodicy, what is needed is a theological inquiry that takes the evolutionary function of affliction at face value and seeks to reflect on its workings in view of the God of Love made known in revelation.

There is a second line of thought, admirable in its own way, which, *2)* rather than trying to fit suffering and death into a logical system ordained by God, addresses them as evils to be fought against. This stance is absolutely right as a human response to distress that can be alleviated. To assuage suffering and promote life's flourishing is a moral imperative in the Jewish and Christian traditions, an expression of love of neighbor deeply intertwined with love of God. In this book, however, we are concerned with the physical process that brings forth the diversity of life. In dialogue with Darwin, we are reflecting only on the second big bang, the natural world in its evolutionary life and death *prior to* and apart from the presence of human beings. This evolving reality needs to be seen and interpreted on its own terms. In this I differ from an important alternative position, as the following examples show.

While agreeing that suffering is intrinsic to the biosphere, British biologist and theologian Celia Deane-Drummond thoughtfully argues in *Christ and Evolution* that to say suffering is necessary, as Peacocke does, is to court the danger of *justifying* it. Thinking of suffering as required, built into the universe in a way that cannot be contested, tends to soften its horror. Better to use Reinhold Niebuhr's account of sin as "unnecessary, but inevitable" to interpret evolutionary suffering. This will guard against any tendency to ignore pain and will motivate us to alleviate it. She poignantly asks, "Why should we help suffering animals on the way to extinction if such processes are part of cruciform existence that cannot ultimately be changed or challenged?"[11] We need to address suffering in a way that gives us a moral imperative to seek its amelioration, not reconcile us with it.

Deane-Drummond is totally right in her view that human beings should act responsibly and with care toward other species, our kin in the community of life. In my view, however, her argument conflates two issues. The first is ethical. The advent of human beings, the third big bang, indeed introduces a moral component when the suffering of other species is caused by human acts. The current disastrous pressure of human behavior on the habitats of other species with the resulting meteoric rise in extinctions of plant and animal species requires a vigorous response of protective care. For the first time in its long history the future of evolution on this planet lies in human hands, and nothing can justify the way we are currently wiping out other species. The need to respond to such unwarranted killing indeed has the character of a moral imperative, and such a mandate should be shouted from the rooftops. There is, however, a second issue, not ethical but biological, which calls for a different response. Convincing evidence exists that pain, suffering, and death existed long before *homo sapiens* emerged and that such afflictions have played an irreplaceable role in the emergence of complex and beautiful life forms. What humans are morally obligated to do for other species does not enter into a theological assessment of this reality. Human beings themselves suffer and die as part of the tree of life that historically developed in this way. Take humans out of the picture, and pain, suffering, and death will continue unabated for other species. That is the issue Darwin's work presents. It needs to be addressed on its own terms.

In his bracing work on *The Spirit of Life*, Jürgen Moltmann agrees that death is the natural end of a mortal life. Countering modern society's denial of death and its efforts to keep it at bay, he argues wisely that to be holy, human beings need to accept the vulnerability and frailty of their lives and integrate a sense of mortality into their self-knowledge and behavior. At the same time, he uses death as a metaphor and more than a metaphor for unjust social and political forces that people inspired by the Holy Spirit must fight against: war, poverty, the systematic destruction of creation.[12] Both positions, personal death as a biological reality that must be maturely integrated and social death as a disaster that we must ethically resist, are well stated. Confusion occurs, however, when the author makes claims that do not distinguish the two:

> The living God and death are irreconcilable antitheses. The hope of resurrection is part of the seeking for the kingdom of God, since the abolition of death is an irrelinquishable component of that kingdom. It is the fault of the religion of redemption to come to terms with death, and to expect eternal life only on the other side of death in a heaven of the saved.[13]

This is actually a third way of perceiving death, as something that if accommodated sets our mind on heaven and interferes with the coming of the reign of God here and now. Again, this is true enough in itself. In terms of our reflection on evolution, however, the absolute statement that "the living God and death are irreconcilable antitheses" is less than helpful. We are precisely trying to think theologically about death as part of the *creative* process on this planet. The creating Giver of life has to be part of the picture in some way, or this is a fruitless endeavor.

Every living organism perishes, writes John Haught in *Making Sense of Evolution*, the latest of his excellent books that have enlightened many on this subject. Organisms have to die in order to make room for the next generation; if they did not, creatures would pile up and evolution would grind to a halt. Whole species also pass away; the drama of life on Earth has featured at least five massive extinctions. The universe as a whole will not last indefinitely either. Naturally speaking, then, death has a functional place in the total scheme of life. Having set out this common consensus about the natural world, Haught then makes a puzzling move. It is essential not to tolerate any intelligible place for death and suffering, he argues. The task of Christian theology is to make it clear that death has no understandable place in the total scheme of things. This would give death a legitimacy it does not deserve, and lead us to tolerate it rather than fight against and overcome it. "I will not make sense of death by staying within the confines of a cramped naturalistic worldview."[14] Instead, theology must appeal to God's eschatological victory over death, symbolized in that stunning scene where God will wipe away all tears and death shall be no more, nor mourning nor crying nor pain, "for the former things have passed away" (Rev. 21.4).

It is extraordinarily valuable to bring in the eschatological viewpoint as, using Haught's powerful category of promise, we will do later in this book. But I do not see why this necessarily undermines the task of Christian theology to reflect on death as an intrinsic part of the creative process in the natural world over billions of years when these "former things" have not yet passed away. Like Deane-Drummond, Haught argues that to acknowledge death in this context would weaken the moral fiber of our resistance to death. Again, I think this confuses evil wrought by human deeds, against which we should indeed fight with every ounce of strength, with the occurrence of natural dying, which theology needs to respect, even for human beings. How could we ever fight against and overcome the death of millions of pelican chicks outside the nest, and why would we even want to? Haught may be on firmer ground if his refusal to accord death a place "in the total scheme of things" is actually a refusal to engage in a theodicy project. It is doubtful however, that this is what he intends, his argument being mounted with reference to natural history.

My purpose in reviewing a recurring bone of contention with these major thinkers and others who could be adduced is in no way to diminish their own contributions, which are significant, but to get clarity on the issue this chapter is addressing. Pain, suffering, and death are intrinsic parts of the process of evolution. As such, they are woven into the very fabric of the origin of species, and need to be distinguished from the harm human beings do. Certainly human beings today must bend vigorous effort to preserve and protect the range of living species, a growing number of which are endangered. Absolutely, the One whom Gustavo Gutierrez brilliantly calls "the God of life"[15] opposes oppression of all kinds, including the grinding down of poor people by unjust economic and political systems and also the wreckage of habitats and life-cycles of living species that are other than human. Surely the eschatological promise of fullness of life for all creation imbues Christian thought and behavior with generative hope. Granting these extraordinarily important insights, I still see a question rising up that has not been fully addressed. Pain and death are basic components of the creation of life on Earth, thankfully not the only components, but nevertheless essential to the way evolution plays out. They render the amazing emergence of life tragic in some dimension. How might theology interpret this reality in a way coherent with a view of the world as God's beloved creation that is good, indeed, "very good" (Gen. 1.31)?

Two convictions govern the reflection that lies ahead. The first considers this terrible phenomenon to be the result of the world's autonomous operation. Affliction arose from below, so to speak, rather than being imposed from above by direct divine will. Theologians are wont to call this the "free process" position. Similar to discussions of free will, which is given to human beings by God yet used at times to oppose the divine will, free process in nature works in ways not necessarily always according to divine design. Polkinghorne expresses this graphically in the image that God is not the "puppetmaster"[16] of either human beings or matter. As we have seen in the previous chapters, the Giver of life allows the world to be itself "in that independence which is Love's gift to the one beloved."[17] In its free working evolution brought forth the kind of life that always entails death and, in its later development, pain and suffering. Without giving creation's affliction ultimate meaning, without rooting it in the eternal will of a good and gracious God, without using it as an excuse not to do good, we begin by acknowledging its existence as part of the finite character of the natural world and respect its role in the evolutionary process.

Then the most fundamental move theology can make, in my view, is to affirm the compassionate presence of God in the midst of the shocking enormity of pain and death. The indwelling, empowering Creator Spirit abides amid the agony and loss. God who is love is there, in solidarity with the creatures shot through with pain and finished by death; there, in the godforsaken moment, as only the Giver of life can be, with the promise of something more. A rich source for this idea lies in the prophetic understanding of the Holy One of Israel as a God of immense pathos who freely relates to the world in delight and anguish. The prophets evoke this divine presence in the midst of catastrophe in unforgettable images. God weeps at the outbreak of war when the harvest of fruit and grain are ruined: "I drench you with my tears" (Isa. 16.9); divine wailing cries out in grief when people are devastated: "my heart moans for Moab like a flute" (Jer. 48.36). In compassion and vulnerable love the Holy One takes up the cry of lament for beloved people who are broken and land that is devastated.[18]

When Christian theology engages this subject, it draws on a peculiar source of insight all its own, namely, the story of Jesus Christ. The experience of a tortured, unjust, tormented death of the worst sort dragged Jesus of Nazareth through godforsakenness into the silence of the tomb. There he was met not by annihilation but by the creative power of the Spirit who transformed his defeat into unimaginable new life in the glory of God. This event, remembered at the heart of every eucharist and celebrated at the high point of the church's liturgical year at Easter, is too rich ever to be completely explained. One thing it does do, however, is inscribe into history the Christian form of hope for redemption. In this telling, the living God redeems the world not by the divine *fiat* of a kindly, distant onlooker but by freely participating in the groaning of the flesh. In Christ, the living God who creates and empowers the evolutionary world also enters the fray, personally drinking the cup of suffering and going down into the nothingness of death, to transform it from within. Hope springs from this divine presence amid the turmoil.

With the entangled bank in view, this chapter attends to the cost of the origin of species in view of the cross. Our sense of the mystery of God's involvement with the world deepens as we ask the beasts about their pain and death in the light of Christ.

DEEP INCARNATION

"It began with an encounter," in Edward Schillebeeckx's eloquent phrase.[19] The carpenter Jesus of Nazareth, a first-century Jewish man in Roman-occupied Palestine, appeared briefly in the public eye, calling disciples to join him in an itinerant ministry that lasted for the span of one to three years. The memory of his teaching and dealings with people, passed on by the disciples and cherished by the communities that came to believe in him, took written shape decades later in the form of diverse gospels. As proclamations of good news, these texts present a riveting picture of a vibrant person, passionately in love with God, who emphasized divine care for all people, especially those who were lost, poor, disparaged, and on the margins of society. "Salvation is on its way from God," he declared with joyous urgency, or, more exactly in the language of his day, "the kingdom of God is at hand" (Mk 1.14).[20] His ministry ended abruptly with his arrest, execution, and burial. Peculiar to Christianity is the fact that what would normally be the end of the story turned unexpectedly into a new beginning as the community of disciples proclaimed that he was risen from the dead, an act of the Spirit which anchors hope of a blessed future for all the world. The arc of this narrative is essential for subsequent christological reflection, and to it we shall shortly return. In order to show the deepest point of connection with the groaning of creation, however, we start with what chronologically came later as a result, the identification of Jesus, crucified and risen, as in person Emmanuel, God-with-us.

In the light of Easter the disciples began to see that Jesus, who had preached that the kingdom of God was at hand, had himself embodied the ways of this reign in an intensely original way. Casting about to find expressions for this startling insight, they pressed into service different metaphors and figures gleaned from the Jewish scriptures: he was Messiah, Son of David, Son of God, Son of Man, Wisdom, Word. The connection they forged between Jesus and Wisdom was especially fruitful in that it began to identify the crucified prophet from Nazareth, localized in time and place, with a divine figure associated in Jewish tradition with creating and governing the world and nurturing human beings on the path of truth and life. Reveling in the world at its beginning, knowing its secrets, indwelling its creatures with her loving spirit, nourishing all with her food and drink, and prevailing over evil, personified Wisdom is one way of figuring the creative, revealing, and saving presence of God in engagement with the world. "In the light of the resurrection, the early Christian community saw Jesus as the Wisdom of God come to us,"[21] writes Denis Edwards, noting how biblical language about Wisdom (*sophia*) is closely related to language about God's Word (*logos*). Each gave the early church a vocabulary for articulating the unique significance of Jesus in relation to God and themselves.

A high point of this developing christology appears in the opening passage of the gospel of John. Like an overture playing before the curtain rises, this poetic prologue sets forth major themes that will appear in the gospel that follows. At its center is the crucial identification of Jesus, the central actor in the subsequent narrative, with God's own creative and saving Word. "Interpreters of John have exhausted every conceivable possibility in an effort to understand the background, meaning, and implications of the Greek word *logos*," writes Smith,[22] referring to sources of "word" in the Greek philosophical tradition, the prophetic tradition of "the word of the Lord" in the Jewish scriptures, and the Christian use of "word" to signify the good news of the gospel, among other possibilities. Interpretation gains more precision when virtually every commentator also notes the close parallels of the Word in this prologue with the story of personified Wisdom in the great wisdom passages of the Bible.[23] With artistic allusions to the creative and saving activity of Wisdom, the prologue narrates the advent of Jesus as the coming of God's personal self-expressing Word, full of loving-kindness and faithfulness, into the world.

"In the beginning," the prologue opens (Jn 1.1). The alert reader hears the first words of Genesis, establishing a link between the story of Jesus and the story of creation. In the beginning before there was a world there was the Word, who was with God, and who was God. Through this Word all things were made, no exceptions. In the Word was life, which was the light of all people; the light shines in darkness, which cannot overcome it. It shines in the world, empowering all who respond to be born as children of God, though some reject the offer. Having established the divine creating and saving character of the Word, the prologue reaches its radical high point with the bold assertion, "And the Word became flesh and lived among us" (Jn 1.14).

Read at the liturgy every Christmas day, this text expresses what would come to be called the doctrine of incarnation (Latin *carne*, flesh), the belief that the living God who is utterly beyond comprehension has joined the flesh of earth in one particular human being of one time and place. The infinite mystery of life and light "lived among us;" the verb here is also ably translated as "dwelt" from the Greek *eskenosen*, meaning to pitch a tent, as in the tent of meeting in the wilderness where God's presence dwelt (Exod. 33.7-11). The personal self-utterance of God within the Trinity, as later doctrine would interpret this, expressed outwards in creation as the Word by which God makes the world, now pitches a tent in the midst of the world, becoming personally part of its history. Jesus dwells among us as the Wisdom of God incarnate, the Word of God made flesh. Henceforth, as Bultmann presses, the glory of God is not to be seen alongside flesh, or through flesh as through a window, but in the flesh and nowhere else.[24]

Note that the prologue does not say that the Word who existed before creation became a human being (Greek *anthropos*), or a man (Gr. *aner*), but flesh (Gr. *sarx*), a broader reality. *Sarx* in the New Testament has multiple meanings. It can have negative connotations, signifying the world as sinful, selfish, opposed to the spirit. In other contexts it simply conveys the finite quality of the material world which is fragile, vulnerable, perishable, transitory, the opposite of divinity clothed in majesty. All emphasis in this gospel text is on the entry of the Word who is God into this mortal realm of earthly existence. Barnabas Lindars underscores what is at stake here with the comment that a reader with the dualistic worldview of Hellenistic thought would be horrified at John 1.14's affirmation that the divine Word became flesh. Far better to hold the docetic view that in Jesus the Word only *appeared* to be human. Put into its historical context, however, the anti-docetic tone of this hymn is unmistakable.[25] It protests against the idea that in Christ the Word of God just made an appearance while remaining untouched by the "contamination" of matter. Taking the ancient theme of God's dwelling among the people of Israel a step further, it affirms that in a new and saving event the Word *became* flesh, entered into the sphere of the material to shed light on all from within.

In truth, the type of *sarx* that the Word became was precisely human flesh. *Homo sapiens*, however, does not stand alone but is part of an interconnected whole. Scientific knowledge today is repositioning the human species as intrinsic part of the evolutionary network of life on our planet, which in turn is a part of the solar system, which itself came into being as a later chapter of cosmic history. The landscape of our imagination expands when we realize that human connection to nature is so deep that we can no longer completely define human identity without including the great sweep of cosmic development and our shared biological ancestry with all organisms in the community of life. We evolved relationally; we exist symbiotically; our existence depends on interaction with the rest of the natural world. Relocating anthropology in this broader context provides the condition to rethink the scope and significance of the incarnation in an ecological direction. The flesh that the Word of God became as a human being is part of the vast body of the cosmos.

The phrase "deep incarnation," coined by Niels Gregersen, is starting to be used in christology to signify this radical divine reach through human flesh all the way down into the very tissue of biological existence with its growth and decay, joined with the wider processes of evolving nature that beget and sustain life.[26] From the beginning God had the character of being a friend of the material world in its full scope, he observes, creating matter, appreciating that it is good, and even declaring that human beings made of the dust of the earth and divine breath were the image of God. Now incarnation enacts a radical embodiment whereby the Word/Wisdom of God joins the material world, sharing in the conditions of the flesh in order to accomplish a new level of union between Creator and creature. The early church axiom that "what is not assumed is not redeemed" carried the insight that it is essential for the divine self-embodiment in Jesus Christ to encompass all that belongs to the creaturely *human* condition, or else it is not saved. Deep incarnation extends this view to include all flesh. In the incarnation Jesus, the self-expressing Wisdom of God, conjoined the material conditions of all biological life forms (grasses and trees), and experienced the pain common to sensitive creatures (sparrows and seals). The flesh assumed in Jesus Christ connects with all humanity, all biological life, all soil, the whole matrix of the material universe down to its very roots.

One can argue this view from the logic of the prologue. It also arises from today's scientific understanding of the world. In becoming flesh the transcendent Word of God lays hold of matter in the form of a human being, a species in which matter has become conscious of itself and deliberately purposive. This matter emerged from the history of the cosmos and is not detachable from the history of the living world. "Born of a woman and the Hebrew gene pool,"[27] the Word of God's embodied self became a creature of Earth, a complex unit of minerals and fluids, an item in the carbon, oxygen, and nitrogen cycles, a moment in the biological evolution of this planet. Like all human beings, Jesus carried within himself "the signature of the supernovas and the geology and life history of the Earth."[28] The atoms comprising his body were once part of other creatures. The genetic structure of the cells in his body were kin to the flowers, the fish, the whole community of life that descended from common ancestors in the ancient seas. "Deep incarnation" understands John 1.14 to be saying that the *sarx* which the Word of God became not only weds Jesus to other human beings in the species; it also reaches beyond us to join him to the whole biological world of living creatures and the cosmic dust of which they are composed. The incarnation is a cosmic event.   NB

Viewing Jesus as God-with-us in this way entails a belief not at all self-evident for monotheistic faith which Christians share with Jewish and Muslim traditions. It affirms the radical notion that the one transcendent God who creates and empowers the world freely chooses to join this world in the flesh, so that it becomes a part of God's own divine story forever. Rahner asserts this truth bluntly, leaving no wiggle room: "The statement of God's *Incarnation* – of God's becoming *material* – is the most basic statement of Christology."[29] It is instructive to watch Rahner wax eloquent against the erroneous but secretly widespread idea that Jesus' human nature was not truly real, that it was no more than a disguise, a suit of clothes which could be shrugged off, a puppet pulled by divine strings, a uniform donned while a certain job is being done, a masquerade in borrowed plumes, an exterior material wrapped around his divine core.[30] This error results, he believes, from assuming that divine nearness is so overwhelming that it will swallow up or at least diminish what is created and finite, whereas precisely the opposite is the case. The non-competitive model of relationship between God who is Love and the world that we have already traced is operative also in the incarnation. Given that nearness to God and genuine human autonomy grow in direct and not inverse proportion, Jesus is genuinely human in virtue of existing as God's own self-expression in time, not despite this. The point in terms of divine relation to the world is that through the incarnation the incomprehensible mystery of God acquires a genuine human life, a story in time, even a death, and does so as a participant in the history of life on our planet. Hence Rahner argues, "the climax of salvation history is not the detachment from earth of the human being as spirit in order to come to God, but the descending and irreversible entrance of God into the world, the coming of the divine *logos* in the flesh, the taking on of matter so that it itself becomes a permanent reality of God."[31]

Given the interconnected character of the material world, the Christ event ramifies throughout the whole creation so that matter in all of its finitude and perishing is fundamentally blessed by being united to God in a new way. Pope John Paul II explained this succinctly:

> The Incarnation of God the Son signifies the taking up into unity with God not only of human nature, but in this human nature, in a sense, of everything that is 'flesh': the whole of humanity, the entire visible and material world. The Incarnation, then, also has a cosmic significance, a cosmic dimension. The 'first-born of all creation,' becoming incarnate in the individual humanity of Christ, unites himself in some way with the entire reality of humanity – which is also 'flesh' – and in this reality with all 'flesh,' with the whole of creation.[32]

Focusing on the human species in itself, the Second Vatican Council spelled out the honor that has accrued: "Since human nature as he assumed it was not annulled, by that very fact it has been raised up to a divine dignity in our respect too."[33] Making the same point more broadly, Teilhard de Chardin praised Christ for "the simple concrete act of your redemptive immersion in matter," drawing out the consequence in a lyrical *Hymn to Matter*. Harsh, perilous, mighty, universal, impenetrable, and mortal though this material stuff be, "I acclaim you as the divine *milieu*, charged with creative power, as the ocean stirred by the Spirit, as the clay moulded and infused with life by the incarnate Word."[34] The incarnation, a densely specific expression of the love of God already poured out in creation, confers a new form of nearness to God on the whole of earthly reality in its corporal and material dimensions, on all of Earth's creatures, on the plants and animals, and on the cosmos in which planet Earth dynamically exists.

In the Christian perspective, the one ineffable God who creates the world is free enough to participate in the created world this way, and loving enough to want to do so. God's own self-expressive Word personally joins the biological world as a member of the human race, and via this perch on the tree of life enters into solidarity with the whole biophysical cosmos of which human beings are a part. This deep incarnation of God within the biotic community of life forges a new kind of union, one with different emphasis from the empowering communion created by the indwelling Creator Spirit. This is a union in the flesh.

THE CHRISTIC PARADIGM

Once Jesus is identified as God's own self-expressing Word in the flesh, the gospel accounts of his life acquire a profound revelatory function. To adapt a felicitous expression from Gregersen, if this is God, then thus is God.[35] If Jesus is God-with-us, then his words and deeds carry a precious disclosure about how incomprehensible holy Mystery, whom no one has seen or can see, relates to the world. Granted the limits of his historical era, geographical location, culture, gender, ethnicity, class, and every other particular characteristic that necessarily mark an individual life, his story inscribes in time a revelation of the heart of God.

Reading the gospel narratives with this understanding sets the divine will for the flourishing of all people into bold relief. Concrete vignettes of Jesus' teaching and characteristic behavior center around the rich Jewish symbol of the reign of God, a biblical expression for the very nature of the indescribable holy One, evoking the moment when the loving power of God will win through over powers that destroy. In parables and beatitudes, healings and festive meals, some of which were scandalous and rife with conflict, this Spirit-blessed prophet provided a joyous foretaste of what the arrival of God's reign would entail.

Workers who arrive toward the end of the day get the same pay as those who labored for hours in the hot sun? A wastrel son is welcomed home with a feast rather than made to pay off his debt? Yes, because God is generous. The religious authorities are full of criticism because Jesus receives tax collectors and sinners and even eats with them? Yes, because God is merciful, like a man who went looking for a lost sheep, and a woman who searched vigorously for a lost coin, both rejoicing when the stray was found. Blessed are the poor and the peacemakers; woe to the selfish rich; the last shall be first and the first last; those who lead should not lord it over others but serve. By means of such parables and colorful pithy sayings, Jesus taught that the compassionate love of God is extravagant, transgressing all cultural and religious expectations of fairness in order to gather in every last hurt or rebellious sufferer.

The good news became ever more concrete when lepers' lesions were closed up; a mentally deranged man was restored to his senses; a hemorrhaging woman had her bleeding stopped; a blind man began to see; a bent-over woman stood up straight with the words, "Woman, you are set free," ringing in her ears (Lk. 13.12). Encounter with Jesus returned suffering people to fuller life in their own finite bodies and made them glad. The meals Jesus hosted or attended as a guest enacted the same good news in a community setting. The wonderful freedom of his inclusive table companionship made experientially real a vital communion with God through conviviality with others. This already offered an experience-in-advance of the reign of God, a world without tears. No wonder joy broke out: "Being sad in Jesus' presence [is] an existential impossibility: his disciples do not fast."[36] Led by the Spirit, Jesus summoned people to conversion, to opening their hearts to the God of mercy and loving-kindness, which in turn impelled them to go and do likewise, loving the neighbor, the stranger beaten and left by the side of the road, even the enemy, leaving no one out.

Interpreting the gospels with contemporary ecological concern prompts the question of whether the "good news" might include the land and its other-than-human creatures. While it would be anachronistic to attribute to Jesus the mind-set and concerns of contemporary ecologists, he and his disciples had inherited the Jewish creation faith of Israel. Consequently, his proclamation that the reign of God was redemptively near assumed that the natural world would be included in this good news. Near the start of his ministry Jesus stood in the Nazareth synagogue and read a text from the scroll of Isaiah which proclaimed not only good news for the poor and liberty for those oppressed but also a year of favor from the Lord, this last evoking the covenant tradition of sabbath and Jubilee years when the land was allowed to rest and recharge (Lk. 4.16-19). Set within an agrarian culture, his parables were salted with reference to seeds and harvests, wheat and weeds, vineyards and fruit trees, rain and sunsets, sheep and nesting birds. He did not hesitate to speak appreciatively about God's care for the lilies of the field, clothed with splendor, and for the birds of the air, bountifully fed.

For someone subsequently interpreted mainly as a spiritual Savior, it is remarkable how strongly Jesus' characteristic deeds focused on bodily well-being. His healing practices placed people's bodily suffering at the center of concern; he even used his own spittle and warm touch to convey health. And how he fed people! Large numbers on hillsides and smaller groups who were his table companions in homes knew firsthand his desire to nourish hungry bodies as well as thirsting spirits. "This man got personal," describes Lisa Isherwood, "sharing touch, engaging with nature, making strong political statements through the symbolic use of food; both during the 'last supper' and in the feedings of thousands Jesus enacted shared power and the interconnected nature of flourishing."[37] The dualism of later Christian tradition that separated spirit from body and saw bodiliness opposed to the divine was not operative in the ministry of Jesus. The God of his heart was the Creator of heaven and earth, and everything was encompassed in the transformation he envisioned.

How to sum this up? Sallie McFague is one theologian who has drawn the discrete gospel episodes together into a brief phrase she has called the "christic paradigm." This functions as a kind of shorthand that illuminates what the ministry of Jesus was all about, a summary of the kernel of what the gospel stories disclose. In a word, "liberating, healing, and inclusive love is the meaning of it all."[38] Those who believe in Christ make a wager that love as Jesus enfleshed it in a human way reveals the ineffable compassion of God; this love is the meaning encoded at the core of human life and at the heart of the universe itself. Write the signature of the christic paradigm, drawn from gospel mercy, across the evolving world. Then it becomes clear that Jesus' ministry reveals that plenitude of life for all, not just for one species or an elite group in that species but for all, including poor human beings and all living creatures, is God's original and ultimate intent. It is not only souls that are important. Physical bodies, gifted with dignity, also matter to God: all bodies, not only those beautiful and full of life but also those damaged, violated, starving, dying, bodies of humankind and otherkind alike. Jesus' ministry grounds compassion for all the bodies in creation. With evolutionary awareness, we observe the christic paradigm take on an ecological dimension.

THE CROSS AND THE TREE OF LIFE

The price exacted for Jesus' fidelity to his ministry was excruciating. Arrested and handed over to Roman authorities by the Jewish chief priests, he was tortured and executed on a cross, the degrading, agonizing penalty reserved for slaves and non-Roman citizens. In the succinct words of the Nicene creed, "he suffered under Pontius Pilate, was crucified, died, and was buried." No exception to perhaps the only ironclad rule in all of nature, his life ended, and in his case ended terribly, unjustly, bleeding out in a spasm of state violence. Theology will have much more to say, but a base point is this: Jesus shared the fate of all who die, which is every living thing. There is a difference that prevents any smooth comparison in that unlike the workings of the natural world that brings death, for example, to the back-up pelican chick, the manner of his death was not part of an evolutionary process. Rather, it was a contingent event resulting from an expedient decision by political authorities. Far from being the result of a natural process, the crucifixion was historical, unpredictable, unjust, the result of human sin. Given the vagaries of human will, his life could have ended differently, and the freedom with which he engaged his death is supremely important for theological interpretation. All these circumstances being noted, still, he died. On a given morning he was a living, breathing entity and by nightfall his body was beyond sensation, cooling in a grave.

That the human being Jesus suffered an agonizing death on the cross is a datum of history. That in this event it was God who suffered and died is a datum of faith, a claim made on the basis of the incarnation: if this is God, then thus is God. Here is limned an astonishing new chapter in the Creator-creature relationship, the Word of God's immersion in matter even unto a suffering death. One influential way of connecting the almighty God with this miserable death is via the idea of *kenosis*, self-emptying. In his letter to the Philippians, Paul encourages members of the church to be of the same mind as Christ Jesus, who though he was in the form of God did not regard equality with God something to be clung to, "but emptied himself," taking the form of a slave, being born human, and becoming obedient, even to death on a cross (Phil. 2.7). This tremendous swoop from divine form to crucified human form traces an arc of divine humility. It credits the incomprehensible God with having a seemingly non-godly characteristic, especially when seen against the model of an omnipotent monarch, namely, the ability to be self-emptying, self-limiting, self-offering, vulnerable, self-giving, in a word, creative Love in action.

Having tasted the dregs of rejection and physical agony, the crucified

Christ knows what it means to suffer. In his own body, he knows. Since he is Wisdom incarnate, this knowing is embedded in the very heart of the living God. As Creator of the world and liberating Redeemer of Israel, the biblical God has always had compassionate knowledge of creation's suffering. This becomes clear early on in the Bible when at the burning bush Moses hears the voice of YHWH, focused on the Israelites enslaved in Egypt, saying, "I know what they are suffering" (Exod. 3.7). The verb "know" here refers to an experiential kind of knowing, being the same Hebrew verb used to describe sexual intercourse in Genesis: "the man knew his wife Eve, and she conceived" (Gen. 4.1). God knows what creatures are suffering; such knowing is continuously part of the Spirit's indwelling relationship to the world. What is new in view of the cross is divine participation in pain and death from *within* the world of the flesh. Now the incarnate God knows through personal experience, so to speak.

God suffers. A construal of classical christology serves to clarify this point in technical language. The incarnate Word's identity is constituted by two natures, human and divine, joined ontologically in a hypostatic union, or union of the person. The *hypostasis* in this case is the second person of the blessed Trinity. Given the profound union of the two natures in one person, church leaders in the fifth century judged it appropriate to cross linguistic wires and attribute human characteristics to the divine nature and vice-versa, the move known as *communicatio idiomatum*, or exchange of idioms. Hence while God is without origin, Mary the mother of Jesus may be called the mother of God, since she gave birth to the human nature of the one who is in person God. This linguistic turn of phrase is more than mere word-play. It expresses the ontological truth of the incarnation. Similarly, while God is fullness of life beyond suffering, it is right to say that God suffered and died on the cross because the human nature of Jesus who suffered is precisely that of the Word of God. In this classic pattern of christological thought, the ultimate answer to the question of *who* suffered on the cross can only be "the divine person" as the one to whom the human nature belongs. David Burrell traces out the logic of this position:

> The actor, to be sure, is the person of the Word, according to a principle which will be enshrined in the scholastic adage: *actiones sunt suppositorum* – "actions belong to the existing subject." Yet inasmuch as that person is also human in Jesus, that very person – the Word of God – can suffer.[39]

In a more contemporary idiom, Walter Kasper describes how the suffering and death of the cross, that unexpected, unseemly event, is "the unsurpassable self-definition of God."[40] The shocking contrast to what we expect divinity to be reveals the unfathomable depths of God's unconditional love. Far from stripping omnipotence away, the self-emptying, weakness, and suffering are not the expression of a lack but of the fullness of divine freedom in love: God "suffers out of love and by reason of his love, which is the overflow of his being."[41] In fact, thinks Kasper, it requires omnipotence to be able to love like this. Philosophical reasoning in the Hellenistic tradition maintains a strict separation between the all holy God and the debility of suffering. Christological doctrine, by contrast, forges a deep personal connection between God and suffering as part of its witness to the abiding, free love poured out "for us" in Jesus Christ.

Contemporary theology is rich in reflection on the power of the cross to bring benefit to human beings in their agony. One of the most telling insights of Jürgen Moltmann's *The Crucified God* holds up the dying Jesus' experience of abandonment, reflected in his prayer, "My God, my God, why have you forsaken me?," and the horrific loud cry that ended his life (Mt. 27.46, 50). By descending into the very depths of godforsakenness, Moltmann reflects, the incarnate Son of God ensures that no one will ever again have to die godforsaken, for divine presence will be there, in the very godforsakenness.[42] It is as if by inhabiting the inside of the isolating shell of death, Christ crucified brings divine life into closest contact with disaster, setting up a gleam of light for all others who suffer in that same annihilating darkness. It is this action and expression of suffering Love that subverts evil from within the world and brings about salvation. Inhabiting a different theological framework, more analogical than dialectical, Pope Benedict XVI preached the same insight: "God has suffered, and through his Son he suffers with us. This is the summit of his power, that he can suffer with us and for us. In our suffering we are never left alone. God, through his Son, suffered first, and he is close to us in our suffering."[43]

In addition to salvific personal effects, the cross also unleashes political meanings. Liberation theologians underscore the historical circumstance that Jesus died as a victim of state policy enforced to carry out the Roman empire's will to dominate occupied peoples. The political nature of his execution connects this terrible moment with all the violent murders unleashed by the power of the state through the ages. This links Christ in solidarity with all whose lives are similarly vanquished, triggering an ethical mandate for the church to resist such injustice, be passionate for peace, bring its mission to the trenches to stop the violence. Crosses keep on being set up in history. The words of Pilate pointing to the thorn-crowned Jesus continue to echo: *Ecce homo*, behold the human being, with the tear-stained, starving, tormented face. Behold the crucified people, in Ignacio Ellacuría's brave analysis, who must be taken down from the cross.[44]

Is the suffering solidarity of the crucified God limited to human beings? Or does it extend to the whole community of life of which human beings are a part? The logic of deep incarnation gives a strong warrant for extending divine solidarity from the cross into the groan of suffering and the silence of death of all creation. All creatures come to an end; those with nervous systems know pain and suffering. Jesus' anguished end places him among this company. The ineffable compassion of God revealed by the cross embraces all who are perishing, not disdaining them in their distress. "Understood in this way," writes Gregersen, "the death of Christ becomes an icon of God's redemptive co-suffering with all sentient life as well as the victims of social competition."[45] Calvary graphically illuminates the insight that the God of love whose presence continuously sustains and empowers the origin of species is a God of suffering love in solidarity with all creatures' living and dying through endless millennia of evolution, from the extinction of species to every sparrow that falls to the ground. As Peacocke carefully contends:

> If Jesus is indeed the self-expression of God in a human person, then the tragedy of his actual human life can be seen as a drawing back of the curtain to unveil a God suffering in and with the sufferings of created humanity and so, by extension, with those of all creation, since humanity is an embedded, evolved part of it.[46]

The cross of Christ concentrates the suffering of God into a point of intensity and transparency that reveals this to be characteristic of God's perennial relation to creation. Dwelling in the evolving world and acting in, with and under its natural processes, the Giver of life continuously knows and bears the cost of new life.

Engaged in the traditional devotion of meditating on the seven last words of Jesus strung together from the four gospels, Arlen Gray came to see that there was an eighth word, the terrible loud cry, which was the last sound out of his mouth. Her dramatic reflection voices biblical and doctrinal teaching in compelling language:

> I suddenly understood that in his final death scream Jesus gathered up all of the earth's suffering throughout all time, bound it up and presented it before the heavenly throne, not in reams of words but in a sacred package encompassing the sorrows, the sufferings, the lost dreams of all creation, all peoples, all times, all conditions, and carried it directly to the pulsing, loving heart of the living Trinity, where it is now. Jesus screams, and he, full of grace and truth, thereby took his and all anguish and transfigured it into a means of touching God.[47]

One may well ask if the presence of the living God with creatures in their suffering makes any difference. In one sense it does not. Death goes on as before, destroying the individual. Wrestling intensely with this problem, Christopher Southgate first admits as much: "When I consider the starving pelican chick, or the impala hobbled by a mother cheetah so that her cubs can learn to pull a prey animal down, I cannot pretend that God's presence as the 'heart' of the world takes the pain of the experience away; I cannot pretend that the suffering may not destroy the creature's consciousness, before death claims it. That is the power of suffering ..."[48] Reflecting further in the light of the cross and resurrection, however, his thought arrives at an awesome insight: "I can only suppose that God's suffering presence is just that, presence, of the most profoundly attentive and loving sort, a solidarity that at some deep level takes away the aloneness of the suffering creature's experience."[49] Admittedly, he grants, this is an anthropomorphic guess. Without psychologizing the chick's or the impala's experience, however, in my view this is one of the most significant things theology can say. Seemingly absent, the Giver of life is silently present with all creatures in their pain and dying. They remain connected to the living God despite what is happening; in fact, in the depths of what is happening. The indwelling, empowering Spirit of God, the Spirit of the crucified Christ, who companions creatures in their individual lives and long-range evolution, does not abandon them in the moment of trial. The cross gives warrant for locating the compassion of God right at the center of the affliction. The pelican chick does not die alone.

DEEP RESURRECTION

The gospel story does not end at the tomb. Led by Mary Magdalen, the women disciples had not abandoned Jesus in his suffering but kept vigil at the cross and accompanied his corpse to the grave. Returning after the sabbath to complete their ritual anointing, they found his body missing. The gospel narratives fracture here, telling disparate stories about the discovery of the empty tomb, angelic messages, and encounters with the at-first-unrecognized risen Jesus by different disciples in the garden, on the road, in the upper room, on the lake shore, on a hill outside the city. The disciples' faith in the God of life took a quantum leap. By the power of the Spirit the reign of God had indeed come and established a beach-head, completely unexpected, in the kingdom of death. Their commission was to witness to this victory.

What resurrection means in the concrete is not seriously imaginable to us who still live within the time-space grid of our known universe. It certainly does not mean that Jesus' corpse was resuscitated to resume life in our present state of biological existence, along the lines of the Lazarus story. Such naive physicality, presented in stained glass windows and Easter cards, pervades popular thinking but it does not bear up under critical scrutiny. Yet the resurrection does have much to do with physicality. The empty tomb stands as a historical marker for the love of God, stronger than death, which can act with a power that transfigures biological existence itself. Theology tends to use the language of transformation at this point but, as Anthony Kelly ruefully observes, "the problem with transformation is that we cannot imagine what it means before it happens."[50] As a seed is unrecognizable in the mature plant into which it sprouts; as the bodies of the sun and stars differ significantly from earthly bodies; as what is perishable becomes imperishable; as a creature of dust comes to bear the image of heaven; as those who are asleep in the grave suddenly become startlingly awake—to cite some of Paul's examples—so too with the new life of the crucified.[51] The angel, a streak of lightening in the tomb, says simply: "he has been raised" (Mt. 28.6).

Starting with a humiliated body laid in a tomb, the resurrection narrative tells of the creative power of divine love "triumphing over the crucifying power of evil and the burying power of death."[52] The Easter message means that Jesus did not die into nothingness, but into the embracing arms of the ineffable God who gives life. What awaited him was not ultimate annihilation but a homecoming into God's mystery. For Jesus personally, this means the abiding, redeemed validity of his human historical existence in God's presence forever.[53]

Glad as this makes the believing community, the *Alleluias* that break out at Easter are based on more than joy for Jesus' personal good fortune. They well up from the realization that Christ's destiny is not meant for himself alone, but for the whole human race. He is "the firstborn from the dead," sings an early Christian hymn (Col. 1.18), signaling that his final destiny also awaits all who go down into the grave, pending judgment. Paul illumines this with the metaphor of the harvest: as the first tomato to be picked hold the promise of the abundance to follow, Christ, "the first fruits of those who have died" (1 Cor. 15.20), gives assurance of a blessed future to those of us still hanging on the vine. In view of the solidarity of the human race, his destiny means that our hope does not merely clutch at a possibility but stands on an irrevocable ground of what has already transpired in him. Life in all fullness awaits. Unimaginable as it may be, this means that salvation is not the escape of the human spirit from an existence embedded in matter, but resurrection of the body, the whole body-person, dust and breath together.

Drawing out Gregersen's insight into deep incarnation that unites the crucified Christ with all creatures in their suffering, I suggest we employ the idea of "deep resurrection" to extend the risen Christ's affiliation to the whole natural world. "The risen Jesus," as Brian Robinette contends, "is in no way extracted from the world's corporeality and history." On the contrary, in a hidden, gracious way, the risen Christ "is to be found at the very heart of creation as the concrete and effective promise that creation is indeed going somewhere." This would not be the case if Easter marked simply the *spiritual* survival of the crucified one after death. But he rose again in his *body*, and lives united with the flesh forever. Herein lies the hinge of hope for all physical beings. In the risen Christ, by an act of infinite mercy and fidelity, "the eternal God has assumed the corporeality of the world into the heart of divine life – not just for time but for eternity."[54] This marks the beginning of the redemption of the whole physical cosmos. With this realization Ambrose of Milan could preach, "In Christ's resurrection the earth itself arose."[55]

The reasoning runs like this. This person, Jesus of Nazareth, was composed of star stuff and earth stuff; his life formed a genuine part of the historical and biological community of Earth; his body existed in a network of relationships drawing from and extending to the whole physical universe. If in death this "piece of this world, real to the core,"[56] as Rahner phrases it, surrendered his life in love and is now forever with God in glory, then this signals embryonically the final beginning of redemptive glorification not just for other human beings but for all flesh, all material beings, every creature that passes through death. The evolving world of life, all of matter in its endless permutations, will not be left behind but will likewise be transfigured by the resurrecting action of the Creator Spirit. The tomb's emptiness signals this cosmic realism. The same early Christian hymn that recognizes Christ as "firstborn of the dead" also names him "the firstborn of all creation" (Col. 1.15).

Christ is the firstborn of all the dead of Darwin's tree of life.

In a beautiful synergy of visual and verbal poetry, the liturgy of the Easter vigil symbolizes this hope with cosmic symbols of darkness and new fire, light spreading from candle to candle, and earthy symbols of flowers and greenery, water and oil, bread and wine. At a climactic moment the *Exsultet*, sung once a year on this night, shouts: "Exult, all creation, around God's throne," for Jesus Christ is risen! The proclamation continues:

Rejoice, O earth, in shining splendor,
radiant in the brightness of your King!
Christ has conquered! Glory fills you!
Darkness vanishes forever!

The moment passes quickly, but it is nevertheless stunning. At the most magnificent liturgy of the year, the church is singing to the Earth! It, too, needs to hear the good news, because the risen Christ embodies the ultimate hope of all creation. The coming final transformation of history will be the salvation of everything, including the groaning community of life, brought into communion with the loving power of the God of life. Because God who creates and empowers the evolutionary world also joins the fray, personally drinking the cup of suffering and going down into the nothingness of death, affliction even at its worst does not have the last word. Hope against hope springs from divine presence amid the death.

With an eye on the entangled bank, we have been pondering the pain and death woven into the very fabric of its evolutionary history. Biologically speaking, organic life's long struggle is always on the way to life forms that are more complex and beautiful. Over thousands of millennia new life comes from death. But the cost is terrible. Pondering the gospel story of Jesus Christ with its consequent doctrines of incarnation and redemption, theology finds a point of strong connection between this affliction and the God of love. Indwelling the world and empowering its creative ways, the ineffable living God also freely joins the world and drinks the cup of suffering, even unto death. Looking back and ahead from the cross, theology can posit divine presence to all suffering and dying creatures, an infinitely compassionate presence that accompanies them knowingly in their pain. What John Paul II calls "the pain of God in Christ crucified"[57] places the living God in solidarity with all creatures that suffer in the struggle of life's evolution. This unfathomable divine presence means they are not alone but accompanied in their anguish and dying with a love that does not snap off just because they are in trouble. Biologically speaking, new life continuously comes from death, over time. Theologically speaking, the cross gives grounds to hope that the presence of the living God in the midst of pain bears creation forward with an unimaginable promise. This does not solve the problem of suffering in a neat systematic way. It does make a supreme difference in what might come next.

8

Bearer of Great Promise

The Spirit of God is a life that bestows life,
root of world-tree and wind in its boughs.
Scrubbing out sin, she rubs oil into wounds.
She is glistening life, alluring all praise,
all-awakening, all-resurrecting.

Hildegard of Bingen

BOOKENDS

In one of her revelations about the courteous love of God,
the fourteenth-century mystical theologian Julian of Norwich
contemplates Christ crucified, seeing his sacrifice as a sign of
his love. Everything which is good and comforting for our help
flows from this love. It is like clothing which wraps and enfolds
us, embraces and shelters us, surrounds and never deserts us. Her
vision of love's embrace then expands beyond human beings to
encompass all the world:

> And in this he showed me something small, no bigger than a
> hazelnut, lying in the palm of my hand, as it seemed to me,
> and it was as round as a ball. I looked at it with the eye of my
> understanding and thought: What can this be? I was amazed
> that it could last, for I thought that because of its littleness it

would suddenly have fallen into nothing. And I was answered in my understanding: It lasts and always will, because God loves it; and thus everything has being through the love of God.[1]

In this chapter we reflect on the hazelnut of the evolving world from the point of view of its beginning, "everything has being through the love of God," and end, "it lasts and always will because God loves it." Having considered *creatio continua*, the abiding presence of the Creator Spirit that continuously empowers the evolving world in its extraordinary fecundity and perpetual perishing, and having linked this to God's unexpected solidarity with the world in the flesh of Jesus Christ, we step back to take a more telescopic view. Where this world comes from and where it is going are fascinating questions. Science rightly answers them, as far as possible, within the framework of time and space. Theology presses these questions to an ultimate point, turning attention beyond time to creation in the beginning (*creatio originalis*) and new creation at the end (*creatio nova*).

Since no one was present to witness the original creation and no one has an advanced script of the final re-creation, both refer to moments that are not open to direct human observation. For theology to say something meaningful at all, it must rest its words on a basis other than hunches and flights of fancy. Such a basis is the living tradition's knowledge of God's graciousness given through Jesus Christ in the power of the Spirit. Borne up by the conviction that God is faithful, the community's understanding can then be predicated backward and forward beyond time, to where no experience can go. On the strength of the experience of grace here and now, the language of faith can claim: as the living God is now, so God was and will be. This way of interpreting faith assertions about the world's ultimate beginning and end makes clear that these are not claims to empirical information but articulations of deep trusting faith.[2]

It was this line of thinking that led the Jewish people to think back from their experience of God's liberating deeds in the Exodus to envision that it was the same God who was powerful enough in the beginning to make heaven and earth. Belief in creation followed the covenant, as biblical scholars point out. A similar pattern of thought led from memory of the Exodus to the later prophetic expectation that God's redeeming action would once again restore the people miserably exiled in Babylon.

A definitive event of revelation shines its light not only backwards but also forward in time in the strong belief that God will still be there, faithful and true.

Adapting this same pattern of interpretation, Christian theology makes protological and eschatological assertions of its own (Greek *eschaton*, the furthest end). Anchored in Christ, the life of the church in the Spirit offers ongoing experience of a good and compassionate God amid the community's own sinfulness and graced commitments. Proclaimed in word and sacrament, experienced in ordinary and extraordinary moments alike, the merciful presence of God, which grasps us at times even in the ache of its absence, gives grounds for speaking with gratitude of an original beginning and with hope of a blessed future. Considerations of the world's ultimate origin and final end launch the mind toward the unknowable. For theology this is the deep mystery of the living God who bears us up in the present.

While talk of beginning and end can make sense if connected with a hermeneutic of the present experience of grace, there is also an inner logic that ties these two ultimate points together. I emphasize this because it seems to have been easier in the course of Christian theology to make a case for the God of the beginning than for the God of the end. Talk of future fulfillment can sound like wishful thinking; biblical apocalyptic descriptions of the end, if not interpreted rightly, can seem like science fiction fantasies. The unreality of it all can be a stumbling block for faith. But there is one God, burning fire of divine love. The logic of belief holds that if this absolute holy Mystery can create life, then this same holy Mystery in faithful love can rescue it from final nothingness.

In scripture the compelling connection between first and renewed creation resounds in the unforgettable words of a Jewish mother during the persecution under Antiochus at the time of the Maccabees. Encouraging her seven sons who were being sequentially tortured to death, she recalls them as babes in her womb, reminding them:

> It was not I who gave you life and breath, nor I who set in order the elements within each of you. Therefore the Creator of the world, who shaped the beginning of humankind and devised the origin of all things, will in his mercy give life and breath back to you again, since you now forget yourselves for the sake of his laws.
>
> (2 Macc. 7.22-3)

And to her youngest:

> I beg you, my child, to look at the heaven and the earth and see every-
> thing that is in them, and recognize that God did not make them out
> of things that existed. And in the same way the human race came into
> being. Do not fear this butcher, but prove worthy of your brothers.
> Accept death, so that in God's mercy I may get you back again along
> with your brothers.
>
> (2 Macc. 7.27-9)

From the Creator who "devised the origin of all things," she reasoned to
God's ability to "give life and breath back" again in the end. Certainly she
assumed that divine power was strong enough to do this. What makes her
testimony so arresting is her confidence in divine mercy, in the compassionate love of God that would *want* to do this, and would.

Two centuries later the apostle Paul again captured the intrinsic link
between divine creative power at the beginning and at the end, this time in
view of the resurrection of Jesus Christ. The God in whom Abraham believed,
he writes, is a God "who gives life to the dead and calls into being the things
that do not exist" (Rom. 4.17). Scholars comment that this is practically a
definition of the Christian God. To give life in the first place and to renew life
that has died in the second place are facets of one and the same divine love
encompassing the world with the same creative power. Hence hope for the
final fulfillment of the world "is nothing other than faith in the Creator with
its eyes turned towards the future," writes Moltmann, succinctly summing up
the internal coherence of creation and eschaton. "Anyone who believes in the
God who created being out of nothing, also believes in the God who gives life
to the dead ... and hopes for the new creation of heaven and earth."[3] If God
can be thought creator, then being thought re-creator of the earth in a transfiguration yet to be realized is not that far a stretch.

With the entangled bank in view we ask the beasts about these transcendent book-ends to the volumes of their story of life.

"WE ARE CREATED"

"We are created:" so the beasts, birds, plants, and fish tell us. What this
means in terms of *creatio originalis* is that they do not explain or ground

themselves, but are brought into being by God's creative word. Closely related to the affirmation of continuous creation, belief in original creation drills down through the abiding presence of the Giver of life to its point of origin. The existence of all creatures is an unowed gift. They exist in a relationship of radical ontological dependence on the overflowing Wellspring of life. And it is good.

The two creation stories that open the book of Genesis teach this truth in the genre of dramatic myth. In the first God speaks and things come forth, starting with the architecture of sky, waters, and land and followed by what Aquinas calls the "adornment"[4] of these spaces with sun, moon, and stars, plants and animals, and the human male and female made in the image and likeness of God. Seeing it all as "very good" (Gen. 1.31) God rests in delight on the sabbath, the feast day of creation's completion and blessing.[5] In the second account of creation that follows in the next chapter of Genesis both the sequence and the manner of creating are changed. First God works like a sculptor to shape a human earth creature from the dust of the ground and breathes up its nostrils the breath of life. God then plants a garden, forms animals and birds out of the same ground, and lastly through a bit of deft surgery introduces sexual differentiation between male and female.[6] The second account, colorfully folkloric, is taken by biblical scholars to be much older than the first, which itself reflects the liturgical hand of its post-exilic priestly author. For all their differences, however, the religious point of both narratives is the same. God alone is the Maker of heaven and earth and all that is in them. No intermediaries are involved. Powerful beings like the sun or certain animals are not deities as neighboring peoples thought, but are creatures, along with everything else.

Creation understood in this way is an original Jewish belief. Biblical commentators note that apart from some few interpretations of creation in view of Christ, such belief receives little elaboration in the New Testament. This is the case, scholars theorize, not because early Christians did not believe the world was created but because that insight had an assured place in their belief system. That God made the heavens and the earth was an integral part of the religious heritage shared by Jesus and his original band of followers; their scriptures were the books of the Hebrew Bible starting with Genesis; there were no controversies around this topic in the early decades after the resurrection. Consequently, creation formed a peaceful part of the religious horizon of the early church. In subsequent centuries, however, the need for further precision arose when the missionary spread of Christianity brought its views into contact with philosophies of the wider Greco-Roman world. It was then that the expression "out of nothing" (Latin, *ex nihilo*), as in God created the world out of nothing, came into use to differentiate the way Christians envisioned the work of creation from two major opposing views.[7]

According to one persuasion, matter was eternal or at least existed prior to the present world; it was the stuff God worked on to create the world. To the contrary, the "out of nothing" phrase signals that there was no pre-existent material, no pantheistic co-eternity of God and matter. God creates by dint of the sheer loving dynamism welling up from the unfathomable plenitude of divine being. To state the point more clearly, "out of nothing" means that God creates but not *from* anything else. Matter is not a presupposition of divine creative activity but a result of it. When humans make, produce, build, shape, generate, or in any way bring something new into being, the action presupposes something already there to be worked on. God's creative act, by contrast, presupposes nothing except the power of divine love which brings into existence something to be worked on in the first place. "Out of nothing" affirms there is only one source of all that is, namely, infinite holy Mystery.

A second problematic tradition encountered by Christianity was Gnostic philosophy which located matter at the lowest reach of reality emanating in descending levels from the divine source, making bodiliness in some way opposed to the spirit which alone is eternally good. In the Manichean version matter itself was brought into being by an evil power, Satan; this rendered the world a site of conflict between the spiritual world of light and the material world of darkness. Even in its less toxic forms, Gnostic philosophy spawned an anti-cosmic spirituality that disparaged matter and encouraged ascent to the realm of spirit, away from the material order of everyday existence. Against the dualism of this view in all its varieties, the phrase "out of nothing" signals the goodness of all things, including material creatures. Since God is the sole source of matter, it is good precisely in its materiality. Yes, the world is genuinely different from the incomprehensible mystery of God, being mortal, temporal, finite. But it is not antithetical to the divine, not demonic. Matter itself is good creation, loved by God in all its otherness.

In the course of time the phrase "out of nothing" came to underscore yet a third insight, namely, that the existence of the world is a free gift. There was no pressure on God to create the world, so to speak. Nothing was there to bring any coercion to bear. Hence creation comes into being not out of necessity but as an absolutely free and unconditioned act of God's own gracious, loving will. The natural world itself is contingent in the sense that it did not have to exist. Its being there at all is dependent on the overflowing generosity of holy Mystery. The statement that God creates from nothing underscores the marvelous incomprehensiblility of this act, allowing God to be truly God. Brian Robinette describes this with careful accuracy:

> Far from making the origin and ground of creation accessible to full comprehension, the statement requires the work of an apophatic discourse that opens up human understanding to the utter gratuity of creation. Nothing is necessary about creation at all. It derives wholly from the incomprehensible mystery of the creator God whose relationship to creation remains one of loving freedom and fidelity. Rather than implying an agonistic picture that situates God and creation in a relationship of rivalry – such a picture only underwrites the serialization of binary and hierarchically arranged terms (e.g., power/weakness, higher/lower, spirit/body, male/female, active/passive, etc.) – *creatio ex nihilo* in fact ruptures such a picture as it emphatically denies that God is "part" of any continuum whatsoever.[8]

The doctrine of creation posits a relationship of origin for all creatures. Their existence in and of itself is totally owed to the free act of an incomprehensibly loving Other. It stands to reason that this belief does not aim to supplant scientific explanation. Far from talking about natural processes which science properly investigates, talk about creation is a type of religious language that refers to the world's ultimate rootedness in a loving power beyond itself. In our day the long twentieth-century debate over whether the universe began explosively at a certain moment or fluctuated in a steady state seems to be resolved in favor of the former view, with evidence pointing to the Big Bang dated around 13.7 billion years ago. But the reality of creation would hold true in either case, since whatever the manner in which the present world came into being, it would still be ontologically grounded in God's creative act.

Creation in the beginning cannot be disconnected from continuous creation. Like air that remains bright while the sun shines, like wood that burns while touched by fire, the world continues so long as God's creative power enlivens it. When a builder puts up a house and departs, reflects Augustine, the structure remains standing despite the fact that its builder is no longer there. But the universe would pass away in the twinkling of an eye if the creating God were to withdraw. "For the power and might of the Creator ... makes every creature abide; and if this power ever ceased to govern creatures, their essences would pass away and all nature would perish."[9] Held in being, continuously created, the world exists from the beginning in a relation of radical reliance on the free gift of a loving God.

The evolving world of species whose dynamic origins Darwin so carefully traced is no exception. Plants and animals exist precisely in this ontological relation of dependence on the Creator. Emerging from the physical conditions of a young planet, life had a beginning in time, now dated between 3.5 and 3.8 billion years ago. It evolved through the interplay of chance and law, some of these laws themselves coming into existence as life emerged and became more complex. Life's explosion into species with their distinctive beauty and nimble interrelationships with other species and the environment finds its ultimate ground in the God of love as creating Source. Their very existence depends on this.

The biblical scene of God's covenant with living creatures adds a measure of divine affection to what might seem a noble but abstract affirmation about sheer gratuity and ontological dependence. Coming at the end of the terrible folkloric story of the flood that wipes out most of life, it signals a change of heart: God decides that never again will living creatures be treated in this way, no matter how badly people behave. When Noah and his boatload of creatures disembark, God says: "As for me, I am establishing my covenant with you and your descendants after you, and with every living creature that is with you, the birds, the domestic animals, and every animal of the earth with you, as many as came out of the ark" (Gen. 9.9-10). With artistic finesse God chooses the rainbow which curves from the ground to the sky and back again as "a sign of the covenant between me and the earth" (v. 13). The divine covenant pledge encompasses non-human creatures in their own right: "When the bow is in the clouds, I will see it and remember the everlasting covenant between God and every living creature of all flesh that is on the earth" (v. 16). Unlike other covenants in the history of Israel, this one does not require affirmation from the creatures; it is a pure gift of assurance and blessing: "This is the sign of the covenant that I have established between me and all flesh that is on the earth" (v. 17). Allowing the religious intuition of this story to shed light on the entangled bank, we can see that for vast tracks of time before humans appeared, and hence without need for human mediation, evolving species on Earth were gifted with existence and held fast over the abyss of nothingness by the creative embrace of divine covenant love. To understand the natural world as creation is to believe that this continues to be the case.

"WE ARE FINITE AND WILL END"

To be created is to be finite and mortal. Such limitation is not evil but simply the condition of being a creature. Nothing on Earth lasts forever. As the evolutionary world of species had a beginning in time, so too it will have an end. Coherent with the idea of creatures' origin in God asserted in *creatio originalis*, the doctrine of *creatio nova* affirms that this earthly end is not ultimately final, leading to annihilation forever. In an unimaginable way the absolute mystery who is *Alpha*, creative God of the beginning, is also *Omega*, the same creative God of the end, who will transform the original creation into "a new heaven and a new earth" (Rev. 21.1), yes, even "a new heaven and a new earth where justice dwells" (2 Pet. 3.13). A blessed future beckons.

To consider this eschatological bookend aright, it is important that scientific and theological language be distinguished. According to contemporary scientific theory the universe itself will come to an end, though uncertainty persists as to the scenario. One option envisions the end happening billions of years hence in a massive fiery contraction. The expanding universe, growing ever larger since the Big Bang, will eventually slow down due to the pull of gravity, brake to a halt, and reverse course. Sometimes called the Big Crunch, this script would end with an implosion, all matter and energy compressed back to a white-hot point of unimaginable density. Some scientists speculate that this would be the point of origin of another explosion like the Big Bang that would set a new universe on its run. No one knows. In a very different scenario the universe's expansion will continue and even speed up, as it seems to be doing in our day, due to elusive forces scientists have named dark matter and dark energy. In this case galaxies will grow further apart; the density needed to create new stars will grow thin; old stars will sputter out without any new ones to take their place. Dark, cold, and silent, the universe will gradually fade to the black of an endless void.[10] One variation of this picture has dark energy pulsing so strongly that it rips matter apart in all directions, destroying even basic atoms. Current calculations favor some version of this latter scenario of cold darkness.

Long before any of these finales take place in the universe as a whole, the evolution of species on planet Earth will have ended. Our sun, a mid-size star, is currently about half-way through its life cycle estimated to be ten billion years, more or less. In about 5 billion years it will run out of its hydrogen fuel and, following the pattern of such stars, will swell up into a red giant. Engulfing the nearby planets Mercury and Venus, the sun's hot expansion will burn up Earth's land and water and blow its atmosphere into space. All life here will die. The planet will be turned into a lifeless cinder circling a star which itself will eventually peter out into a cool dwarf. Some scientists speculate that by that far future human beings will have figured out how to migrate to other compatible solar systems, bringing our plants and animals with us. Perhaps. Even if we do, that solar system, too, will come to an end, as will this universe itself. For the sake of this discussion, I want to remain realistically on planet Earth with its evolving origin of species adapted to the environment here. Without a doubt, this gorgeous community of life has a roughly datable end.

The theological assertion that on "the last day" the cosmos will be renewed in a new, transfigured life with God does not deny any of these scenarios. Its claim is based not on the potential of the finite world in itself to survive the final death, but on the character of God. As we have seen, such hopeful language arises from the religious community's experience of God's gracious love and plays it forward. It expresses trust that the same God of steadfast love, made known through the history of Israel and in Jesus Christ and tasted in contemporary experiences of the Spirit, will still be present and active in the future. In this framework, whatever theology says taps into the conviction that the God of love is faithful, and will hold fast to what is fragile and finite.

Hope like this is far from an easy optimism that overlooks or underplays what is negative. It faces the end full on: at one time in the past there was no life on earth, and at some time in the future there will again be no life on the shriveled earth. To hope for more in clear-eyed view of the end of life on our planet entails a certain hanging on in the dark, "hoping against hope" (Rom. 4.18), as scripture would have it. This phrase comes from the story of childless Abraham who saw his hundred-year-old body as good as dead and his old wife Sarah's womb utterly barren, yet still believed in God's promise that he would be the father of many nations. In an analogous way, "hoping against hope" aptly expresses the wager being made by belief in a renewed heaven and earth. Indeed, it is extreme circumstances like the death of a beloved person or of the living planet itself that put into high relief the nature of religious language about the future as the language of daring, abiding trust in the fidelity of God.

What this blessed future of the species will look like is impossible to imagine. It is worth lingering on this point. Remember that we are not making predictions or engaging in some sort of futurology. Eschatological language is shot through with reverent reserve, with a "deliberate unknowing" which refuses to claim that any particular scenario is confidently to hand.[11] What such speech does do is affirm the core conviction that all of reality exists within the embrace of God's gracious love, and that it is going toward a fulfillment yet to come. Hence the various apocalyptic passages in scripture should not be read as if they are giving advance information about a coming historical chain of events. Like the creation stories in the book of Genesis, these end-of-time texts aim to reveal a key element about the world, namely, that its destiny is totally in the powerfully loving hands of God. Whether terrifying (chaos will erupt, stars will fall) or comforting (a heavenly wedding banquet, God will wipe away all tears), the poetic images employed are not forecasts to be interpreted with wooden literalism but narrative ways of teaching religious truth. The Giver of life who created all beings out of nothing will still be there after final devastation, holding fast to the beloved creation. Without knowing particulars, these texts essentially declare the hope that the ultimate future will be blessed. Nothing more, but also nothing less.

COSMIC REDEMPTION

Ecological awareness in our age is bringing back into theological focus this biblical hope for the whole world's redemption, or put in other words, its transformed fullness of life. Readers may find such an idea initially strange, since connecting the saving work of Christ with creatures other than human has not been a predominant move of the western theological tradition. Common understanding that "Jesus died to save us from our sins" would seem to preclude the natural world being affected, since plants and animals do not sin. To be clear, the New Testament does emphasize the powerful healing effects of Christ's cross and resurrection on human beings held in the grip of multiple oppressions. We are justified, forgiven, declared no longer guilty in the eyes of God, freed for a life of love, justice, and peace. As Paul tellingly writes, "There is therefore now no condemnation for those who are in Christ Jesus" (Rom. 8.1). This splendid good news, however, was never intended to exclude the rest of creation.

Silence on this matter became especially pronounced after Anselm's satisfaction theory took hold in the 11th century in the West. Humans have sinned and offended the honor of the infinite God, the theory begins. We cannot make suitable atonement, being finite creatures. The death of the innocent Son of God on the cross is the event that makes abundant satisfaction for sin to the just holiness of God; in mercy this satisfaction is shared with all who are sinners, and we are saved. In its original form reflecting its feudal context, the intent of the satisfaction theory was to spell out divine mercy, for the initiative to become human and die in order to save the human race is all God's, and the resulting satisfaction made by Christ is shared with all who are sinners. In the hands of lesser preachers over time, however, the theory came across as though God's anger over sin needed to be appeased by a bloody death, and as though Christ's sacrifice was necessary to win divine favor. In subsequent centuries theology used this theory to interpret the cross in a largely legal, juridical way; penitential practices in the church put this interpretation into vigorous practice. It is worth noting that in this construal of the cross, neither the ministry of Jesus nor the resurrection of the crucified play a significant role in the good news of salvation. Obviously, too, the natural world is not in the picture. The focus is narrowed down to human sin and the cross.

Beginning in the mid-twentieth century theology started to explore ideas of redemption that were prevalent prior to Anselm with a view toward renewing the meaning of the doctrine for contemporary believers. Two sources of refreshment were the scriptures and the theology of the first Christian centuries in the East which continues in the Orthodox churches today. Examining the full range of New Testament writings, scholars have found richly diverse ways of understanding salvation for humans beyond the metaphors of atonement, satisfaction, and sacrifice. While the mystery of grace poured out in the crucified and risen Jesus remains central, the meaning of this gift is expressed in what Schillebeeckx calls different "interpretive elements" throughout the New Testament. Biblical writers elaborated the good news using concepts of liberation, reconciliation, healing, justification, victory over the powers, living in peace, fullness of life, being freed from slavery, adoption, and new birth as God's children, to name but a few.[12] These long-untapped resources, coupled with the importance of Jesus' ministry and the restoration of the resurrection to the paschal mystery, open doors to understanding more varied dimensions of what is meant by the mystery of redemption.

One result has been renewed awareness of New Testament texts about cosmic redemption that previously just flew by. These texts that extend the promise of a future to all of creation are few in number, but they are strong. We have already considered Paul's reflection in Romans 8.18-25 with its powerful sense of creation's groaning and its affirmation that creation will be set free from its bondage to decay. The fourfold use of "the creation" in this passage intertwined with language about creation's sharing in the freedom of the children of God clearly signals that the destinies of human beings and of all creatures without exception are laced together. In other writers the phrase "all things" (Greek, *ta panta*) raises the same flag of inclusiveness. The great hymn in Colossians which draws on the Wisdom tradition and the history of Jesus in equal measure, is suffused with this insight:

> He is the image of the invisible God, the firstborn of all creation; for in him all things in heaven and on earth were created, things visible and invisible, whether thrones or dominations or rulers or powers – all things have been created through him and for him. He himself is before

all things, and in him all things hold together. He is the head of the body, the church; he is the firstborn from the dead, so that he might come to have first place in everything. For in him all the fullness of God was pleased to dwell, and through him God was please to reconcile to himself all things, whether on earth or in heaven, making peace by the blood of his cross.

(Col. 1.15-20)

The drumbeat of "all things" repeated five times in this short text, coupled with reference to "all creation," "everything," and the encompassing "things visible and invisible," drives home the blessing that flows to the whole world from the cross.[13] Similarly, while praising the riches of grace lavished on the saints, i.e. everyone who is in Christ, the letter to the Ephesians reveals the mystery of God's good pleasure, a plan for the fullness of time which is to gather up "all things" in Christ, things in heaven and things on earth (1.10). The visionary writer of the book of Revelation hears "every creature in heaven and on earth and under the earth and in the sea, and all that is in them" singing praises to the Lamb (Rev. 5.13), and perceives a climactic event of transformation where the One who sits on the throne says, "See, I am making all things new" (21.5). With its roots in the Abrahamic and prophetic traditions and its own singular revelatory experience of the risen Christ, the New Testament includes a hope-filled vision of the whole universe pervaded with divine promise.

In seeking to refresh its understanding of the meaning of redemption as fullness of life for all, Western theology receives a rich assist from another living tradition. Without denying the need to be redeemed from sin, the Eastern church's theology has long focused more strongly on the need to be delivered from *death* and its corruption. Consequently, its theologies of redemption have always had a more cosmic scope, encompassing all finite reality that tastes death. Rooted in scripture, developed by Greek patristic theologians, and enthusiastically expressed in liturgical celebrations of Easter, this tradition remains alive in Eastern Orthodoxy to this day. Kallistos Ware describes its commitment clearly:

creation is to be saved and glorified ... This idea of cosmic redemption is based, like the orthodox doctrine of icons, upon a right understanding of the Incarnation: Christ took flesh – something from the material order

– and so has made possible the redemption and metamorphosis of all creation – not merely the immaterial but the physical.[14]

In dialogue with Orthodox tradition, western theology is growing in its ability to do justice to the long-overlooked tradition of the redemption of the whole biophysical world.

An illuminating instance of how these developments have coalesced in new theological insight is found in Karl Rahner's discussion of the cross. From the outset, he begins, the world and its history are based on God's will to bestow the free gift of grace in radical abundance upon the creature. Reaching its climax in the historically tangible event of incarnation, this divine will to self-communication already surrounds sin with an offer of forgiveness. Hence, the popular notion attendant upon the satisfaction theory that Jesus' death effects a sea-change in God's attitude from anger to forgiveness is simply not tenable. The God who is love does not need to be so persuaded. Let us say without verbal subterfuge, Rahner urges, that with regard to mercy, God's mind cannot be changed. Let us also maintain that in the matter of salvation all initiative proceeds from God's own will. Then in what sense can we say that Jesus' death brings about salvation? By making graphically real in history the love of God who joins us in death in order to bring life out of that state of nullity. Rahner finds this to be "a causality of a quasi-sacramental and real-symbolic nature." God loves us first, and then does something to rivet this saving love forever into the history of the world. "In this causality what is signified, in this case God's salvific will, posits the sign, in this case the death of Jesus along with his resurrection, and in and through the sign it causes what is signified."[15] The point being that the self-communicating love that brought the world into existence to begin with has now enacted that love in a historical life climaxing in a moment of terrible tragedy and unfathomable victory that can never be reversed. This brings about a new realization of grace with concomitant liberation from sin and death. To say that Christ saves us from sin is to highlight only one aspect of the mystery: "properly speaking it is not Christ's action which causes God's will to forgiveness, but vice versa."[16]

Undergirding Rahner's interpretation of the cross is the alliance he made with the medieval Franciscan position associated with Duns Scotus.[17] In Aquinas and most others, discussion of the motive for the incarnation took its cue from the Genesis narrative, to wit: the world was created good; our first parents sinned, ruining their relationship with God; therefore, in mercy the Son of God became incarnate and died in order to restore the relationship. The Scotistic position, by contrast, maintained that the incarnation would have taken place whether human beings had sinned or not. Why? The chain of reasoning of this alternative position has to do with the dynamic of love. God is unfathomable love; love seeks union with the beloved; this union occurs in the incarnation when the divine Word enters into personal union with the created world in Jesus Christ. Even if Adam and Eve and their descendants were still innocently in the Garden, this would have happened; it is the way love acts. As a matter of fact, however, human beings did sin, setting up the antagonistic conditions in which Jesus' life ended in suffering and agonizing death. In this way divine love showed itself capable of going all the way into the depths of degradation in order to heal from within. But the incarnation is not dependent on the sin of our first parents. It was Love's very intent from the beginning.

This medieval interpretation of the Christ event, predicated on the love of God taken with extreme seriousness, opens out gracefully to an interpretation of the cross that underscores more than an atoning sacrifice for sin. To be sure, the death of Jesus is indissolubly connected with sin. It is an epitome of the evil humans do to each other for state power to take a healthy man and reduce him to a bloody corpse through torture and violence. Scotus among other pre- and post-Anselmian readings of this horrific event, however, locates its redeeming power not in satisfaction rendered to a God whose honor has been violated, but in the presence of divine love in the flesh enacting an historical solidarity with all who suffer and die.

This line of thinking prompts Rahner's further reflection on cosmic redemption. The narrow juridical orientation characteristic of much traditional Latin soteriology, he contends, omits the transfiguration of creation which belongs so organically to resurrection hope. Let us understand that Jesus was a child of earth. As such he died, descending into the dark bosom of the grave. The fact that he died does not mean that his soul freed itself from the world and fled into the immensity of God's glory beyond the murderous earth which shattered his body. No. Jesus rose again in his body. Here the living God has done something new: called the flesh to life again in transformed glory. Risen from the dead, Jesus "has been born again as a child of the earth, but of the transfigured, liberated earth, the earth which in him is eternally confirmed and eternally redeemed from death and futility."[18] The tomb may be empty, but far from departing the earth the risen Christ remains in its midst as the first exemplar of the radiance that awaits all beings:

> His resurrection is like the first eruption of a volcano which shows that in the interior of the world God's fire is already burning, and this will bring everything to blessed ardor in its light. He has risen to show that this has already begun.[19]

Easter proclaims a beginning which has already decided the remotest future for all. We know that the natural world is in motion with its own evolving history, as are human beings. Everything is still on the way. What is the destination? A bottomless void of nothingness or an all-embracing blessedness? In Jesus of Nazareth, crucified and risen, earth's history has already reached its culmination in one of its own. This end has been made manifest to the world still advancing through time, "just as the front of a procession which has reached the goal calls back with cries of triumph to those still marching: we are there, we have found the goal, and it is what we hoped it would be."[20] The living God is the future of the world. Time is still stamped with the suffering of Good Friday; days still wander in the silence and loss of Holy Saturday. Yet in wordless expectation, all earthly reality is in tremendous movement toward its own glory which has already arrived by the power of the Spirit in the risen Christ. Put another way, Christ carries the whole creation toward its destiny. His resurrection is the beginning of the resurrection of all flesh. Or so Christians hope. They hope that when the last day does come, it will be nothing less than "the universal Easter of the cosmos."[21]

Rahner's is but one way to plot a hope-filled future for all creation. Biologically speaking, new life continuously comes from death, over time. Theologically speaking, without diluting the affliction with a facile hope, the language of faith can dare to say that the encompassing mystery enacted in Jesus Christ through the Spirit bears creation forward with an unimaginable promise toward a final fulfillment when God will be "all in all" (1 Cor. 15.28).

MUIR'S BEAR

When the noted U.S. naturalist John Muir came across a dead bear in Yosemite, he penned a bitter complaint in his journal about religious folk whose belief had no room in heaven for such a noble creature: "Not content with taking all of earth, they also claim the celestial country as the only ones who possess the kinds of souls for which that imponderable empire was planned." These magnificent creatures, however, are expressions of God's power "inseparably companioned by love." They are made of the same dust as we, and breathe the same winds and drink of the same waters. A bear's days are warmed by the same sun, and his life, pulsing with a heart like ours, was poured from the same First Fountain. With our stingy spirit we may want to block this creature from heaven. To the contrary, he figured, God's "charity is broad enough for bears."[22]

Muir's complaint is not without basis. It is disconcerting to discover how, despite scriptural affirmation of cosmic redemption, wide swaths of theology traditionally limited how much of the living world would in fact be included in the final blessing of a redeemed cosmos. In his study of nature's travail, Santmire developed the category of symmetry-asymmetry as a tool to clarify where a host of thinkers stand on the issue.[23] A minority view, evident in early post-apostolic writers such as Irenaeus, theologians of the eastern church such as Basil, and singular figures such as the later Augustine, envision that *all* living beings are included in the world's final transfiguration into the new heaven and the new earth. There is *symmetry* in their thought between creation and redemption: God creates all things and God will save all things from final annihilation. While agreeing on creation's goodness, however, a more numerous cadre of thinkers, stretching from Origen and the early Augustine through the medievals Thomas Aquinas and Bonaventure to the Reformers Luther and Calvin and onto Barth and beyond, concentrate on humanity's need for redemption to such a degree that an *asymmetry* creeps in. God creates all things but will not necessarily save all things. The rich vision of the redemption of the whole created world becomes peripheral and dims out into a primarily a-cosmic notion of redemption for human beings alone. The asymmetry grows more pronounced when even for human beings the resurrection of the body becomes a secondary consideration to the rescue of their immortal soul.

In the asymmetric trajectory, theology has more often than not denied a redemptive future for plants and animals both as species and as individuals. Consider the explanation adduced by Aquinas. To the question "Whether the world will be renewed?" he answers most certainly yes, arguing among other positive reasons that "the dwelling should befit the dweller. But the world was made to be the human being's dwelling. Therefore it should befit humans. Now human beings will be renewed. Therefore the world will be likewise."[24] In addition, since "human beings love the whole world naturally and consequently desire its good," the universe will be included in the final renewal to satisfy this desire. Notice, however, that the question does not ask if the *whole* world will be renewed. It turns out that only certain parts will be so blessed. These include the heavenly bodies which will shine more brilliantly, the earth with its elements such as water, air, and fire which will be clothed with a harmonious brightness, and resurrected human beings who will find the transformed heavens and earth a fit dwelling for their glorified bodily selves.

To the explicit question "Whether the plants and animals will remain in this renewal?," however, the answer is decidedly no. Aquinas has already explained that "We believe all corporeal things to have been made for man's sake."[25] These include plants and animals that serve human beings primarily in two ways: first, as sustenance to their bodily life, and secondly, as helping them to know God, inasmuch as humans glimpse the invisible things of God through the visible things that are made. Clearly these two services will not be needed in the glory of heaven. Human beings in an incorruptible state will exist without having to eat, and they will know God directly through the beatific vision. Since plants and animals are no longer needed for these human benefits, they will not be included in the renewal of the world. Just as important for Aquinas' argument, these creatures are not capable of being eschatologically renewed. In matter and form they have not been endowed with anything incorruptible such as the imperishable nature of the shining heavenly bodies or the rational soul of human beings. "Therefore," Aquinas concludes, "plants and animals will altogether cease after the renewal of the world."[26] Created by almighty God who pronounces them good, they come and go, serve their purpose, and ultimately pass into nothingness. Herewith lies the ambiguity of asymmetry: the biophysical world which is created will not ultimately be redeemed in the glory of the last day.

The relish with which Aquinas engaged the new science and natural philosophy of his day gives good reason to expect that he would shift his position on this question in light of the evolutionary knowledge of a later age. As it is, his work stands among legions of others as a particularly clear example of reasoning that sets up a debilitating denial of a redemptive future to the species whose dynamic process of origin Darwin described. What might theology venture in our day? For sure, we have no advance information about the details of life after death, even for human beings. What happens after death is unknown in the concrete: "eye has not seen nor ear heard" what God has in store (1 Cor. 2.9). Christian faith affirms that human persons are brought into new life, enfolded in a communion of love, even to the point of being "face to face" with the unspeakable mystery of the living God. Can an analogous fulfillment be predicated of species and individual organisms which are not human beings? As Denis Edwards asks in an immensely helpful essay, "Do individual sparrows and kangaroos participate in their own way in redemption in Christ?"[27]

In this debate some thinkers such as Rolston propose that insofar as the death of creatures provides nutrients for others to live and the extinction of species provides living space for new forms of life, nature is being continually redeemed. Redemption need not involve new life for an individual creature. The reemergence of life in others is an extraordinary value and recompense enough for any organism's suffering. Viewed theologically, he contends, the way of nature is the way of the cross in which life comes from death. As creatures struggle, they suffer through to something higher; God is working throughout the whole process to bring about a fullness of life.[28] Others such as Southgate argue that this view stops the meaning of redemption too soon. The new generation of another creature is not enough to make things work out all right, especially in face of the unthinkable scope of suffering over millennia. We need to stake a claim, he contends, for the future of species and even the individual organism by introducing in broad eschatological terms God's saving will toward all creation.[29]

To my way of thinking, a case can be made that for God to love the whole means to love every part. Hence to save the whole means to save every individual, every bear. I suggest that this position is based on core truths of faith and coherent with their dynamism. To wit:

⚜ The living God creates and cares for all creatures.

⚜ This love encompasses all creatures even in their suffering and dying.

⚜ These creatures are part of the flesh of the world which the Word of God joined via incarnation.

⚜ The death and resurrection of Jesus offers hope of redemption for all flesh.

⚜ The life-giving presence of the Spirit who empowers all creation is also the power of resurrected life for all beings.

Given the personal presence of divine love to every creature in every moment, and the further revelation of the character of this love in the suffering and hope-filled story of Jesus Christ, there is warrant for holding that species and even individual creatures are not abandoned in death but taken into communion with the living God. Nothing is lost. For human beings and other living organisms as well, the promise of final redemption in both a general and particular sense seems fitting in view of the goodness of God whose love treasures every creature.

How this can be accomplished is beyond what we can figure. Virtually all scholars working on the question include the important reservation that redemptive fulfillment will be appropriate to each creature's own capacities. Reflecting on the hope that the backup pelican chicks left to die through the millennia will find fulfillment in some way beyond death, Jay McDaniel insists colorfully that "If there is a pelican heaven, it is a *pelican* heaven."[30] God relates to each creature on its own terms, and thus its fulfillment will be one that fits its nature. "Based upon God's wisdom and justice," writes Edwards, "I believe that it can be taken as a fundamental principle that redemptive fulfillment of any creature will be specific to the creature involved."[31] Resurrection entails a radically transformed bodiliness, he continues, and many would think the promise of life to come holds true for their beloved dog. They may well be right, but redemption for a mosquito may be of a different order, "one that in God's wisdom is fully appropriate to a mosquito."[32]

As theology probes this idea of hope for the creatures, a certain literal-mindedness can plague and even shut down this discussion. Still, the play of imagination opens the door to religious possibilities, as the following examples from a both preacher and a poet attest. In 1782 John Wesley preached a terrific sermon entitled "The General Deliverance" that remains in print and is still being studied. Keenly aware of the deplorable suffering of animals, especially at the hand of their worst predator, human beings, he preached that because God is merciful and just the "brute creatures" will be saved, brought as individuals to eternal life. Their vigor will be restored, even increased, and they will be beautiful again, thriving in a perennial springtime: "In the new earth, as well as the new heavens, there will be nothing to give pain, but everything that the wisdom and goodness of God can create to give happiness. As a recompense for what they once suffered while under 'the bondage of corruption' ... they shall enjoy happiness suited to their state, without alloy, without interruption, and without end."[33] The details, spun from Wesley's great heart, are imagined, but the conviction they express, that the animals will be blessed, is rooted in wisdom about God's great heart.

The poem "The Heaven of Animals" by James Dickey depicts animal happiness more concretely, including an exquisite image of the predator-prey relation:[34]

> Here they are. The soft eyes open.
> If they have lived in a wood
> It is a wood.
> If they have lived on plains
> It is grass rolling
> Under their feet forever.
>
> Having no souls, they have come,
> Anyway, beyond their knowing.
> Their instincts wholly bloom
> And they rise.
> The soft eyes open.
>
> To match them, the landscape flowers,
> Outdoing, desperately
> Outdoing what is required:
> The richest wood,
> The deepest field.

For some of these,
It could not be the place
It is, without blood.
These hunt, as they have done,
But with claws and teeth grown perfect,

More deadly than they can believe.
They stalk more silently,
And crouch on the limbs of trees,
And their descent
Upon the bright backs of their prey

May take years
In a sovereign floating of joy.
And those that are hunted
Know this as their life,
Their reward: to walk

Under such trees in full knowledge
Of what is in glory above them,
And to feel no fear,
But acceptance, compliance.
Fulfilling themselves without pain

At the cycle's center,
They tremble, they walk
Under the tree,
They fall, they are torn,
They rise, they walk again.

Affirming that the promise of new creation includes all creatures as individuals in a way appropriate to their nature is not a foolish construal. Based on the belief that the Giver of life indwells each creature to empower its life within the evolutionary process, and that the same Spirit of the crucified and risen Christ accompanies each creature in its pain and dying, this position figures it would be discordant with the fibre of such creative love to allow any creature just to disappear. Hope for the creatures finds backing in this teaching of Jesus which evokes divine care for every little bird: "Are not two sparrows sold for a penny? Yet not one of them will fall to the ground apart from your Father" (Mt. 10.29). This saying is part of a larger discourse assuring human beings of God's care for them: "And even the hairs of your head are all counted. So do not be afraid; you are of more value than many sparrows" (10.30-1). While humans are being assured, however, the point not to be missed is that God's caring embrace also extends to every sparrow that dies. This teaching is strengthened in Luke's version of the same saying which shows how cheaply these birds passed hands: "Are not five sparrows sold for two pennies? Yet not one of them is forgotten in God's sight" (12.6); it seems one bird is thrown in for free if you place a double order. The hope that living creatures have a destiny in glorified life with God rolls out an implication of this teaching that they do not die "apart from your Father" and are not "forgotten in God's sight," not only corporately but singly, "not one of them." Honoring the unfathomable measure of this divine love, I think a good case can be made that Edwards is right when he concludes:

> The Creator Spirit is with creatures in their finitude, death, and incompletion, holding each in redemptive love, and is in some way already drawing each into an unforeseeable eschatological future in the divine life. This promise points to an unimaginable participation of all creatures in the dynamism of divine life. The Spirit is with each creature now, with every wild predator and its prey and with every dying creature, as midwife to the unforeseeable birth in which all things will be made new.[35]

Any scientific proof, of course, is out of the question. By its own methods science cannot rightly know any more than at death each individual creature truly dies; that extinction means that a species disappears forever; and that life as a whole on this planet will one day end, most likely a few billion years hence. Theology, a different discipline, acknowledges these facts, all the while believing that death does not have the last word. Even the end of the world can set no limits to the divine Giver of life who created the world to begin with. In hopeful trust in God who is faithful, theology dares to affirm that the living world with all its members is being drawn toward a blessed future, promised but unknown.

There are grounds for thinking that God's charity is indeed broad enough for bears.

In dialogue with Darwin these chapters (5–8) have characterized the evolving world of plant and animal species with significant religious markers. Continuously fired into being by the Giver of life, the living world is the dwelling place of God. Ontologically dependent on the Creator, it is empowered with the autonomy befitting a finite creature to operate freely in the course of its own evolution. In solidarity with the perishing of Christ who shares its flesh, it is a groaning, cruciform world, destined for resurrection. Existing in absolute dependence on its Maker, it bears the promise of new eschatological life, heading toward a final fulfillment thanks to the *Alpha* and *Omega* whose fidelity knows no end. The God-world relationship developed here draws much of its insight from pneumatology and christology. The liberating presence of the Spirit understood as the very love of God proceeding, and the play of the life, death, and resurrection of Jesus Christ understood as Word made flesh, bespeak an intimate nearness of the divine that graces the natural world. To round out this picture, the next chapters bring in the human species. Human beings are not aliens parachuted in from some other world but are natural-born members of the community of life germinated out of the depths of the evolutionary process. In their own way they share all the facets of relationship to the living God described above, with the added note of capacity for profound moral and spiritual responsiveness. Their presence makes a huge difference to the world of life.

9

ENTER THE HUMANS

> They are clouds without water, blown by the wind; autumn
> trees without fruit, uprooted, twice dead; wild waves of the sea,
> casting up the foam of their own shame; wandering stars, for
> whom the deepest darkness has been reserved forever.
>
> Jude 1.12-13

AN EVOLVING SINGULARITY

From a scientific perspective, there is no doubt that human beings evolved as a twig on the branching tree of life. Out of colonies of single-celled creatures in the ancient seas came diverse species of creatures that live in shells, radiating species of fish, amphibians, and reptiles, in tandem with bushes, trees, diverse types of insects and flowers, ramifying species of birds and mammals. The mammalian branch itself sprayed out into various hominid forms, and from one line of descent emerged modern human beings, we primates whose brains are so richly textured that we experience self-reflective consciousness and freedom, or in classical terms, mind and will. This account of origins places the human race in a complex relationship of continuity and discontinuity with the rest of the planet's flora and fauna. Physically, like all living creatures, we emerge from the material universe, being constituted by its elements. The life-fluid in grasses and birds shares with human bodily fluid the common molecular structure of water; eyes respond similarly to electromagnetic waves of light; the rhythms of day and night pattern creaturely behavior. In addition, given the common ancestry posited by evolution, we humans also share a genetic heritage with every other species of the tree of life, a biological kinship encoded in each cell of our bodies. These uninterrupted similarities clearly mark *Homo sapiens* as a species that belongs to planet Earth. At the same time, human cognitive powers and ability to act freely with deliberate intent mark our twig as a singularity, a species with extraordinary capacities never before seen. Philosophers of science capture the similarity to and difference from the rest of life characteristic of the human species with the Zen-like comment that in human beings the universe has become conscious of itself.

In line with the principle that a new species will evolve close in time and space to its immediate ancestor, the major though not undisputed consensus today holds that the genus *Homo* (Latin, human) originated on the African continent. This is to be expected since Africa is where the closest living relatives to humans are to be found. Data derived from DNA sequences reveal that the immediate human chapter begins with a primate ancestral lineage that branched out into gorilla, chimpanzee, and hominid groups. Gorillas split off first. Then approximately 6 to 5 million years ago the chimp and hominid lineages, which share on average more than 95 per cent of their genetic make-up, diverged.

For the next few million years the planet was home to a wide range of related hominid species, most notably the *Australopithecus*, a group of bipedal apes that includes the iconic fossil Lucy from Ethiopia. By two million years ago in Africa, the genus *Homo* made its appearance as one of the diverging sprays of the hominid line (recall Darwin's diagram). The exact time, place, and ancestor of this new shoot remain unknown. In turn the genus *Homo*, human, itself became a branching lineage that produced multiple species which came into being and faded away. Some existed simultaneously. Based on the fossil record and molecular analysis, science has identified various human species, although not without dispute: *H. habilis*, *H. rudolfensis*, *H. ergaster*, *H. erectus*, and *H. antecessor*, among others.[1] Each inherited roughly modern human body proportions yet with changing physical characteristics (size of cranium, position of larynx, shape of pelvis, feet, and hands) and cultural abilities (tool-making, domesticating fire, building shelters), giving evidence of what Camilo Cela-Conde describes as "a gradual and slow evolution of mental processes."[2] Some of these species migrated out of Africa to the Middle East and thence eastward to Asia and Australia and westward to Europe in successive waves. Fossils of another human species, *H. neanderthalensis*, have been found in Europe but not in Africa, indicating that human evolution continued abroad. In the midst of this variety of older human species, the cognitively and anatomically modern humans, *H. sapiens*, emerged in Africa somewhere between 200,000 to 100,000 years ago. Neither the exact moment nor the immediate line of descent is known as yet. This species, too, migrated out of Africa and now covers the planet.

A popular misconception holds that a straight and simple line leads from our common ancestors, the early hominids, to modern human beings. Recall a classic cartoon that depicts a sequence of creatures emerging from the sea and gradually ascending a slope to modern upright human form. Here evolution is seen as a linear progression in which each successive species simply replaces its predecessor. Darwin's diagram of taxa with its little sprays of species, most going extinct, some changing and continuing on through time to give rise to yet more diverging species, presents a scenario closer to the evidence now being discovered. The human lineage resembles a bushy phylogenetic branch. For most of our genus' evolutionary history a diverse array of human species inhabited Earth. Today, however, modern *H. sapiens* is the only species left in its genus. All the cousins have gone extinct.

What propelled the evolution of *Homo sapiens* as a unique species on the hominid branch, according to paleontologists, biologists, and anthropologists, was a remarkably rapid (in evolutionary terms) increase in the size and complexity of the brain, accompanied by changes in the position of the larynx which made language possible.[3] Already matter had organized itself into an array of self-reproducing animals capable of receiving and processing signals from the environment and exercising cognitive awareness and emotional response. Already the bipedalism of human ancestral species had freed the hands to develop fine-grained maneuvers. Already skilled use of stone tools and fire by other human species, along with burial of the dead, had generated a recognizable form of culture. Now a threshold was crossed. The astonishingly beautiful cave art of Ice Age Europe, which dates to between 35,000 to 15,000 years ago, gives evidence that, as Ian Tatersall writes, "the wonderful and unprecedented human creative spirit was already fully formed at that distant point in modern pre-history."[4] What marked this species was self-consciousness, use of language, and tremendous fluidity in behavior.

Consider the human brain, the most complex biological organ in the universe, as far as we know. Folded in on itself like an intricate Chinese puzzle, it has billions of neurons, each with several thousand synapses, making possible the processing and transmitting of information through electrical and chemical signals. Rolston underscores the wonder of this organ with an apt comparison. When envisioned on a cosmic scale humans may be minuscule figures, but on a scale of mental complexity we are immense: "in our 3-pound brain is more operational organization than in the whole of the Andromeda galaxy."[5] The physical and chemical configurations of this organ do not of themselves explain the vast array of intellectual and volitional powers human persons exercise, let alone our experience of self-consciousness. The mind cannot be reduced so easily to the material function of the brain, though in the scientific view its working requires this basis. Rather, as with all evolutionary changes, a new complex organization of matter allowed new capacities to emerge, capacities that require new levels of explanation. How such capacities might be explained, however, is today fiercely contested, the mind-brain relationship being one of the most lively areas of research with positions taken across a wide spectrum of scholarly stances.[6]

Given this kind of brain, human consciousness became rich with self-reflective, symbolic, and linguistic capacities. People do not just register information about their environment like other animals, but in a subjective return to their inner selves can know that they are doing so: I know that I know. People do not just form mental images of the physical world around them, but can imaginatively dissect and reassemble these into "a huge vocabulary of intangible symbols,"[7] which they use to generate explanations of the world, signify meaning, and orient their lives. And people can pass ideas from one mind into another through the medium of language, a deeply mysterious thing we do. As ideas and their practical results are handed on through the generations, cultures take shape. Cultures are shared worldviews, patterns of behavior, and affective understandings that are learned through a process of socialization. The transition into culture, a psycho-social phenomenon, occasions a non-linear, exponential jump in intellectual powers. Ideas can jump across genetic lines: "one does not have to have Plato's genes to be a Platonist, Darwin's genes to be a Darwinian, or Jesus' genes to be a Christian."[8] Henceforth human evolution proceeds by the interplay of two fronts: biological, governed by genetic variation and natural selection, and cultural, a mental and social realm rife with new freedoms.

As a result, human beings find themselves uniquely emplaced on the planet. Bodily earthlings who like all other living creatures interact with their environment as they are born, wax, wane, and die, they are yet able mentally to transcend any particular time and place. They ask questions, dream of what comes next, wonder about the meaning of the whole. Able to act with deliberate intentionality, they choose goals beyond biological survival and reproduction and act to achieve them. They make art in visual, aural, tactile, and literary mediums. And can they ever innovate and invent! A highly ethical animal, they can consider principles of right and wrong, weigh what they ought to do, and chose one path over another in the face of temptation. They can interact as an agent with other similarly free, existential agents and hold each other accountable. They can love with deep emotion, self-giving, and spiritual exhilaration, and hate just as strongly. They can even love an enemy. In view of their singular self-reflective inwardness, cognitive power, and freedom of action, their philosophers describe them as persons composed of body and soul, rational animals, spirited selves, embodied spirits, spirit in the world. Religious teachers add that they are created in the divine image and likeness, being a complex unity whose body comes from the dust of the earth and whose soul is breathed into them by God, each one gifted with unique dignity.

The point is, the human species is a singularity. To varying degrees other species in the animal kingdom make creative use of the environment, communicate with each other, feel emotions, grieve their dead, and may even recognize their own faces. Contemporary studies of living animals are making clear that the gap between humankind and otherkind is much less absolute than previously thought, with many shared characteristics appearing on a graded spectrum. At the same time, *Homo sapiens* is not simply one more sibling. As of this writing, for example, scientific study indicates that the human genome differs from that of chimpanzees, our closest animal relatives, by approximately 4 per cent. But it is the human species alone that has done the work of sequencing the two genomes. Therein lies the difference: "For all the manifold talents that chimpanzees possess, that cognitive gulf still yawns."[9] With *Homo sapiens*, evolution on this planet has brought forth a creature able to decipher the very process of evolution and draw diagrams about its progression. In so doing, it has brought forth a being that can massively effect the evolution of other species, for good or ill.

The extent and quality of this influence now goes beyond anything *On the Origin of Species* envisioned. The human species is having a huge impact on the evolution of the rest of the natural world not simply by practices of selective breeding of animals and plants, but much more significantly by propelling vast numbers of other species toward extinction. Mass extinction results when change is too rapid and large for adaptation to be an option. Humans today are acting as a potent agent of evolution by destroying habitat and changing the environment so rapidly and in so many ways that numerous other creatures cannot keep up. Hence they are disappearing in catastrophic numbers. By any measure our late-arriving species is a marvel. We are incandescent with the power to think and choose. We have advanced capabilities to respond to other beings, to imagine the thought worlds of others, to act out of a sense of moral obligation, to respond aesthetically to the beauty of nature, even to praise the Creator of that beauty. Despite our unique capacities for language, reason, morality, and love, however, the human legacy is becoming the erasure of others on the tree of life.

EAARTH (SIC)

From the beginning the advent of the human species had momentous consequences for other living species and for the flowing systems of air, water, and soil nutrients that make all life possible. Coming out of Africa, this species covered the Earth in a blink of geological time, occupying climate zones from freezing to tropical. A bird's-eye view of its development is astonishing. Starting out with a hunter-gatherer style of life, *Homo sapiens* continuously elaborated new ways of interacting with the natural world:

domesticating plants and animals, taming fire, forging metals into tools, building complex structures, and processing foodstuffs and skins in an ever increasing array of skilled crafts. This enabled human populations to increase, spread, and gather into dense urban concentrations. The onset of the Industrial Revolution in the eighteenth century racheted up human use of natural resources, with machines fired by fossil fuels doing what had been the hard, slow work of people and animals in agriculture, transportation, and production of goods. The proclivity to experiment has since given humans the ability to release power from the nucleus of atoms, strike out for the moon, and communicate by technological wizardry that captures words and images over long distances.

In our day the cumulative effect of this activity on the natural world has reached damaging proportions. For a time Earth could replenish its physical resources after human use, but no longer; we are depleting supplies of clean water and healthy soil at too rapid a rate. For a time other species could largely regrow their populations after human predation, but no longer; our zooming numbers, coupled with habits of consuming and polluting, have coalesced into an engine of destruction for others. The human species has even become a geophysical force capable of raising the planet's temperature: searing droughts, annual "once-in-a-century" floods, mega-fires, massive storms, and rising sea levels give evidence that the weather itself is becoming traumatized.[10] Ecological activist Bill McKibben argues we should respell the name of our planet, rendering it *Eaarth*, to signal that the planet on which humans thrived for 10,000 years, "the sweetest of sweet spots" that enabled successful farming and civilization's great cities to take shape, no longer exists. It may look familiar enough, but since the late 1960s the planet has reached a tipping point toward profound changes with dire consequences for other species. It is not just that the human thumbprint has been impressed on every nook and cranny, but that the very atmosphere has been tampered with, affecting everything it envelops: "Name a major feature of the earth's surface and you'll find massive change."[11] We may not be intending for this to happen, but as a matter of fact it is.

Reams of empirical reports in the media and scientific literature are analyzing what is happening, but for many people these just fly by. To grasp some idea of human impact on the living world, I invite you to perform a characteristic human feat and run a slow-motion video in your imagination.

Start with the planet, alive with interlocking ecological systems from north to south, east to west, land to sea to air. Color these ecosystems different shades of blue. Now observe a trio of human behaviors that are disrupting this envelope of biological life. First, increasing billions of human beings spread out over the globe, pervading the living space of other species. Envision each person as a little brown dot. Next, note where these billions of brown dots consume resources at an exponential rate, the affluent among them leaving slim pickings for the poor of their own species, to say nothing of other species. Mark such patches of depleted resources with the color orange. Finally, add tints and wavy rivers of yellow to the brown-spotted, orange-blemished blue Earth. These represent continuous flows of toxic pollution that degrade the land, sea, and air, the habitats where other species have evolved and now try to live. Let the brown dots, orange patches, and yellow streams emit together a red haze of warmth that rises around the earth. Intensify these colors in view of the following facts.

Population growth
The biblical injunction to "increase and multiply" may be the only one that humankind has obeyed faithfully, jests James Nash.[12] But the result is no laughing matter. From its first appearance sometime in the last one to two hundred thousand years until the year 1650 of our era, the human species grew to comprise around one-half billion members. This number doubled to one billion by the early 1800s, and doubled again by the mid-twentieth century, so that by 1950 more than two billion people were living simul-taneously on the planet. A mere fifty years later that number tripled; at the turn of the millennium in 2000 there were six billion humans. It took only ten years for the count to increase another billion, giving the world population a count of seven billion by 2010. While it took the human species thousands of centuries to produce one billion people, that number is now being reproduced in a decade. Any chart that plots human population growth now ends with a spike that goes up off the page. Predictions vary as to where this growth might top out; perhaps ten billion by mid-twenty-first century, fifteen billion by the twenty-second century?

Common sense can see that this dramatic progression puts intense pressure on other species. People need land to live on; minimally they need food, water, and shelter to survive. Their increasing, skillful use of these resources diminishes the access of other species to these same necessities, squeezing out their living space. As *Homo sapiens* increases exponentially, other species decrease. The overarching reason lies in the fact that while Earth is an enormously rich planet, it is not infinite. Although technologies may extend the ability of certain resources to support life, for example, water-purification systems, in the end Earth has a finite carrying capacity. There are limits.

It may seem that introducing the problematic impact of population growth on the world of other species brings this discussion into the thicket of contentious debate. While the question of *how* to control population growth does indeed divide interested parties at the global and national level, it is important to note that in recent decades the Roman Catholic Church has endorsed the basic idea that it is legitimate to limit human births. Speaking of the responsibility of married couples to determine the number of children they will have, the Second Vatican Council teaches:

> Let them thoughtfully take into account both their own welfare and that of their children, those already born and those which the future may bring. For this accounting they need to reckon with both the material and the spiritual conditions of the times as well as of their state in life. Finally, they should consult the interests of the family group, of temporal society, and of the Church herself.[13]

In this same vein Pope John Paul II, while disavowing the use of artificial contraception, stated in an address that the church "fully approves of the natural regulation of fertility and it approves of responsible parenting." Coining an evocative phrase, "morally correct levels of birth," he continued, "This morally correct level must be established by taking into account not only the good of one's own family, and even the state of health and the means of the couple themselves, but also the good of the society to which they belong, of the Church, and even of all mankind."[14] To these criteria, it would not be surprising in the near future to see explicitly added "the good of all creation," given the recent growth of church teaching about moral responsibility for the well-being of the natural world. If the good of future children, the material conditions of the times, and the interests of society are factors in weighing the ethical rightness of reproductive activity, the good of the ecological world which sustains human society is also profoundly relevant. In this light, one can conclude that not all levels of human birth are morally correct.

Resource Consumption

Spreading human populations have a history of using natural resources to the point where they become depleted. Then if possible they move on, looking for more animals to hunt over the next hill, more fish to catch in the next bay.[15] On a finite, self-contained planet, this cannot continue forever, and we have now reached a point of resource exhaustion. Nonrenewable fossil fuels such as oil, gas, and coal and industrially significant minerals such as iron, copper, and nickel will by definition run out. Of much more significance for other living species are the naturally regenerative resources of whole ecosystems with their topsoil, fresh water, vegetation, and prey species necessary for survival. These are becoming functionally nonrenewable due to abusive human practices such as clear-cut logging of forests, overgrazing grasslands, depleting underground aquifers, and siphoning off river water for agricultural, industrial, and urban purposes. It isn't human use as such but the extent of it that is inflicting deadly damage on the ability of ecosystems' natural cycles to renew themselves. Soil erodes, farmland becomes salty, semi-arid areas dry out into desert. The destruction of coastal wetlands, a principal nursery of sea life, coupled with overfishing by technologically sophisticated fleets, causes populations of targeted fish to crash, to say nothing of the collateral damage to other species such as dolphins and turtles caught in the dragnets. The sea floor becomes a desert. This litany of havoc could continue, with the ongoing effect that living space for other species disappears. There is not an ecologist alive who would disagree with James Nash's judgment: "If sustainability implies living within the bounds of the regenerative capacities of the earth, with a sense of responsibility for future generations, then present practice is characterized predominantly by *un*sustainability in the use of both nonrenewable and renewable resources."[16]

Maldistribution of resources among the human species itself complicates the picture. Excessive consumption by a well-off minority of people goes hand in hand with desperate poverty for starving millions. Economists estimate that 25 per cent of earth's people in the affluent nations annually use roughly 75 per cent of the world's resources. Such consumption of goods in the course of a comfortable life for average persons in affluent nations contributes as much if not more to environmental degradation than overpopulation in poorer nations. Some would say that in terms of per capita resource consumption, the United States is the probably the most over-populated nation in the world.[17]

On a structural level galloping consumption is driven by economic market systems that demand constant growth in order to be viable. Pursuing profit with a commitment comparable to religious fervor, national and global corporations seek a bottom line always in the black, not the red, let alone in the green, allowing the ecological cost of doing business drop from view.[18] Poor people suffer disproportionately from environmental damage inflicted in pursuit of corporate profit. Business practices entailed in mining, logging, oil-extraction, plantation farming, and industrial fishing remove the resources of poorer nations, depleting their ecological richness without commensurate recompense. Ravaging of people and ravaging of the land on which they depend go hand in hand.[19] Corporate logging of forests in India, to take one example, not only ruins the habitat for wildlife but deprives subsistence communities that live on the forests' periphery of the firewood, fruits and nuts, small animals, and clean drinking water they depend on for survival. In the Amazon basin, lack of just distribution of land pushes dispossessed rural peoples to the edge of the rain forest where in order to stay alive they practice slash-and-burn agriculture, in the process destroying pristine habitat, killing rare animals, and displacing indigenous peoples. The brilliance of Wangari Mathaai's Green Belt Movement in Kenya, for which she was awarded the Nobel Peace Prize in 2004, lay in the unity it forged between planting trees, women's empowerment, and a stable democratic peace. Critics questioned what reforestation, let alone the economic and political well-being of poor women in a patriarchal society, had to do with peace. As the movement has spread to other countries, however, it becomes increasingly clear how profoundly smart it is to advocate replenishing the land and empowering neglected persons whose well-being cannot be separated in building a stable society.[20]

With exquisite shortsightedness some theorists and activists have set up a choice between social justice and ecojustice, but this is to miss the larger picture. Ecological integrity and socioeconomic justice intertwine in a tight embrace. The former is an essential condition for the latter, the two forming not parallel conditions but a mutually reinforcing cycle. Yet in nations such as the United States and in the world of nations as a whole, the gap between rich and poor people continues to grow, with corresponding damage to ecological systems. At the same time, largely unnoticed, all around our burgeoning human species there abide countless other species whose lives hang in the balance, subject to our gobbling up resources.

Pollution

The healthy functioning of land, water, and air in their natural states is increasingly befouled by human actions that introduce contaminants. Ecosystems can normally assimilate a certain degree of pollution; they have consistently done so over time, self-regulating back to a productive state when their workings are interrupted by natural disasters. The intensity of human-generated pollution in many places in our day, however, is exceeding the capacity of natural systems to regenerate. The impact on other living species can be disastrous. Consider these defilements: oil spills, toxic discharges into air and water from refineries and other industrial sources, pesticides sprayed over miles of farmland, accidental emissions from chemical and radioactive facilities, seeping wastes from mines, ground-level ozone smog, emissions from motor vehicles and incinerators, concentrated human excrement poured into waterways, and the fouling effect of municipal dumps and landfills (the pollution of profligacy[21]). How can living creatures survive this poisoning of their world? Oil-soaked aquatic birds, collapsing bee colonies, damaged vegetation, tainted fish, and disappearing song birds give mute testimony to pollution's death-dealing effects. Over all these toxicities now lies the effect of climate change, an effect of burning fossil fuels that emits the pollution of excess carbon dioxide into the atmosphere. Rising temperatures are changing the habitats where species thrived for generations, disrupting their food and reproductive cycles.

In one engaging case study, *Heartbeats in the Muck*, John Waldman examines the sea life in New York Harbor before the coming of the Europeans until now, lifting up a famous example. The fantastic abundance of oysters in the harbor estuary had long been a mainstay of the Native American diet. Early explorers registered miles of gigantic oyster reefs, and soon were making oysters into soups, patties, and puddings which were eaten for breakfast, lunch, and dinner. The island where the Statue of Liberty now stands was once known as Oyster Island. In the 18th century one naturalist observed that "there are poor people who live all year long upon nothing but oysters and a little bread."[22] Discarded shells were used for paving streets or burned into a form of lime used in building houses. Stalls along the piers in lower Manhattan and oyster houses close by touted "Rockaways" and "Amboys," named after the regional beds they were taken from. In the nineteenth century one traveler remarked, "Everyone here seems to eat oysters all day long."[23] Gradually the human impact kicked in. Overharvesting and smothering by sewage sludge reduced the number of live oyster reefs. Due to increased pollution, "particular beds were gaining reputations for producing oysters that tasted like petroleum." By the early twentieth century, Waldman reports, "typhoid fever outbreaks from contaminated shellfish from Jamaica and Raritan Bays ended any lingering interest in consuming the remnants of the local oyster stocks."[24] Today a few solitary oysters inhabit crevices here and there among rocks in the harbor. Contamination makes them unfit for eating. The great living reefs are gone.

See now in your mind's eye the variegated blue of the planet's ecosystems swarmed over with increasing billions of brown dots, splotched with orange depleted resources, swirling with yellow effluents in land, sea, and air, surrounded with an increasingly red envelope of warming haze. This is the current situation. Denial does not change the fact that "our planet is sick at the structural level, the level where health is necessary if the planet is to provide the resources for all life-forms to flourish."[25] A crisis can be defined as an emergency, an unstable situation of terrible danger or difficulty, a crucial stage or turning point in the course of something. To say that Eaarth is in the midst of an ecological crisis is no exaggeration.

EXTINCTION: NEVER AGAIN

To complete this imaginative picture of the planet, begin to erase all colors here and there, creating empty spaces in the messed-up blue. These bare streaks and spots stand for the absence of living species that have gone and are going extinct. The main lethal cause is destruction of their physical habitat. The onslaught of our one species is ruining the living places of multitudes of others. Consequently, with their home gone, magnificent animals and tiny flowers that took millions of years to evolve are being forced out of existence. They will never return. No fallout from human action is more devastating to the tree of life.

As we have seen, Darwin's theory holds that extinction plays an essential role in the life-producing process of evolution. Analyzing long millennia of time, biologists estimate that on average the background level of extinction has been about one species per year. This is an average, and the rate differs for different groups of organisms. Small species such as insects, bacteria, and fungi have disappeared at the rate of about 10 to 100 species per year. Mammals go for years without any species disappearing, their background extinction rate in the past being approximately one species lost every 200 years.

Recall that in addition to this infinitesimally slow disappearance of species over millions of years there have probably been five catastrophic events of mass extinction. Scientific consensus today is increasingly of the mind that Earth is on the verge of, or even well into, a sixth mass extinction event. This time, however, death is not being caused by the break-up of continents, a chance asteroid collision, or a chain of naturally-occurring climatic changes. Instead, the cause is the activity of one species, the mushrooming *Homo sapiens.* The first documented case of extinction in modern times was that of the aurochs, a giant type of wild cattle, the last known group of which lived in the Polish Royal Forest west of Warsaw. In 1557 they numbered 50; though considered precious and carefully protected, 40 years later their number was down to 25; the last female died in 1627.[26] Since then, although no single agreed-upon statistic illustrates the damage, evidence gathered from around the world paints the disturbing picture that extinction is proceeding at a rapid rate, far above pre-human levels. In contrast to Earth's normal background rate of extinction which sees one species naturally finishing its life-span every year, 150 to 200 species now become extinct *every day*, according to a 2010 calculation of the UN Environment Programme.[27] Estimates based on the fragmentation and destruction of tropical forests say that we are likely losing 27,000 species per year from those habitats alone. Compared with the average loss of one mammal every two centuries, the count since 1600, when two mammals on average should have disappeared, has been 89 mammalian extinctions, almost 45 times the predicted rate; as of this writing, another 169 mammal species are listed as critically endangered. Wilson articulates a broadly-held conclusion: "Clearly we are in the midst of one of the great extinction spasms of geological history."[28]

In the first decade of the twenty-first century, the list of disappeared creatures includes the golden toad, the black-faced honeycreeper, the Baiji dolphin, the West African black rhino: amphibian, bird, and mammal species that represent thousands of others, large and small, that have vanished from land, sea, and air.[29] Vegetation, too, is affected:

> Everywhere, trees are dying. The boreal forests of Canada and Russia are being devoured by beetles. Drought-tolerant pines are disappearing in Greece. In North Africa, Atlas cedars are shriveling. Wet and dry tropical forests in Asia are collapsing. Australian eucalyptus forests are burning. The Amazon basin has just been hit by two severe droughts. And it's predicted that trees in the American Southwest may be gone by the end of this century.[30]

Current forecasts anticipate that as many as one-quarter to one-third of the world's animals and plants are likely on a path to extinction within the next hundred years, with tropical rainforest species, top carnivores, species with small geographic ranges, and maritime reef species at especially high risk. Despite the pioneering science done by Jane Goodall, Dian Fossey, and Biruté Galdikas on our next-of-kin species, "apes are hurtling toward extinction in the wild. Their forests are being logged and converted to plantations. Gorillas are dying from Ebola. Chimpanzees are hunted for food or as illegal pets. To study apes today is not to discover them for the first time, but perhaps to say farewell."[31] Many species are currently numbered among the "living dead," populations so small that they have little hope of survival. Other species are vulnerable because of their ecological relationships; the loss of a pollinator, for example, can doom the plant it pollinates, and a prey species may take its predator down when it vanishes. Most endangered ecosystems comprise hundreds of interacting species; thus when keystone species are ruined they take batches of others with them into oblivion.

Try to grasp what extinction means. In this event it is not just an individual that dies but a unique configuration of animal or plant. These species are not like stamps or other collectibles; each has survived the long journey through evolutionary stops and starts, and exists with a unique grace in the community of life.[32] Exquisitely adapted to their ecological niche by millions of years of evolution, species now disappear. The disappearance is irreversible. The nature of the evolutionary process which requires biological connection to an immediately ancestor on the tree of life means that these vanished creatures will never again exist on the planet. There is a terrible finality here. The nature of this annihilation comes to expression in the insight, "Death cuts off life; extinction cuts off birth."[33]

A threefold loss ensues: for the species itself, for its potential evolutionary future, and for the strong ecological network of life on this planet. The species itself is gone from the planet, and new members who have not yet been born will never see the light of day. Snap, the great aurochs break off. Crack, there goes the pied raven. Rip, no more Chinese elephant, Bali tiger, Carribean monk seal, or passenger pigeon. Slash, the Saint Helena olive, Rio de Janeiro myrtle, Moorea laurel, Hawaii ruta tree disappear. Furthermore, given the flow of genes from ancestor to future progeny, an unbridgeable break appears. Any new evolutionary emergence from this branch of the tree of life is finished; no new possibility of as yet unimagined species will ever be realized in this lineage. The species and their future evolving descendants are gone forever, never again to contribute their particular grace to the ecological community. Finally, when this wipe-out is multiplied to include thousands of species annually, the resulting loss of biodiversity is ecologically dangerous. It breaks up the envelope of life that surrounds our planet as a whole. The strong assembly of species which took millions of years to evolve causes the Earth to be a richly productive place, hospitable to the thriving life of humankind and otherkind alike. In face of the current human onslaught on ecosystems, however, E. O. Wilson rightly asks the terrible question: "how much force does it take to break the crucible of evolution?"[34]

Several times while writing this book I was asked why the current human-induced extinction of species should be considered so awful, given that the history of life is punctuated with great die-offs due to natural causes. The answer is that what is happening now is not at all natural. The unparalleled scope and appalling pace of extinction in our day is due to a preventable cause. Species are meeting their demise prematurely by assault at the hands of a cognitively powerful, volitionally free fellow species. The appropriate analogy is murder. Similar to the violent killing of human beings in their youth or prime, species that should be alive are being slammed into permanent disappearance by a disastrous failure of human wisdom and will. Rather than allowing their death to come naturally in old age after millions of years of evolution, human action is prematurely shutting species down. We should be holding funerals.

While the fossil record shows that biodiversity has always recovered, it also indicates that recovery is naturally slow, taking 5 to 10 million years after the mass extinctions of the past for an array of new species to evolve. As Wilson explains in the concrete:

> Great biological diversity takes long stretches of geologic time and the accumulation of large reservoirs of unique genes. The richest ecosystems build slowly over millions of years ... only a few new species are poised to move into novel adaptive zones, to create something spectacular and stretch the limits of diversity. A panda or sequoia represents a magnitude of evolution that comes along only rarely. It takes a stroke of luck and long period of probing, experimentation, and failure. Such a creation is part of deep history, and the planet does not have the means nor we the time to see it repeated.[35]

This means that in the case of the current mass extinction, more than 200,000 generations of humankind will have to live and die before levels of biodiversity comparable to those we inherited at the start of the twentieth century might be restored, if ever.

In an insightful essay the novelist Lydia Millet details one effect. In a way unique among species, many human children grow up with animals as companions of their imagination through whom they explore the world. In stories, books, movies, and toys, they find comfort in animals' companionship and moral lessons in their exploits. Then "what of the children of the future? When the polar bears and penguins are gone, the gorillas and elephants and coral-reef clown fish like Nemo – what diverse and lovable army will be their close companions?" We adults may grieve their loss now, but our grandchildren will know of them only by hearsay, as children today know of the dinosaurs. "This planet will no longer be our old, familiar home, but something completely other. And that will change the character, the aesthetics, the ideals of our descendants, growing up on a globe that has almost in the blink of an eye been purged of its ancient evolutionary richness." We'll be sending those children into a starker, poorer land whose many possibilities have been eternally foreclosed. Will the children easily turn their attention to an array of bright novelties, robots and such? "Or will they, every now and then – after watching an old movie or reading an old book and glimpsing the marvelous strangeness and beauty of what once lived here with us – imagine those sad multitudes with dragging wings and drooping tails and make a childish wish: Can't you come back? Come back, come back! Come back to us, you great, dead creatures of the earth."[36]

They will not, of course.

If human beings were to wake up to the grandeur of the living world, fall in love with life, and change their behavior to protect it, much of the current dying-off could be slowly brought under control. But in our day the dire situation appears to be accelerating, with humanity's rapacious habits driving species to extinction faster than new species are able to evolve. The tree of life is thinning out.

THE PROMISE OF NATURE

Why should anyone care? One prevalent line of argument holds that it is in our human self-interest to protect biodiversity. Besides the beauty and comfort that the richness of the natural world provides, the disappearance of myriads of species means that ecosystems are likely to lose much of their ability to render many valuable services that we take for granted, from cleaning and recirculating air and water, to pollinating crops, to providing a source for new pharmaceuticals. We need to leave a living Earth for our children and grandchildren that will continue to make possible their healthy and productive lives, unto the seventh generation. These reasons focused on human benefit are indeed true. Would that they would be more efficacious as spurs to action. The fact that they do not galvanize action to protect other species casts humanity's self-designation as *Homo sapiens*, or the wise human, in a deeply ironic light.

A more loving, less self-interested reason for caring lies in the importance of the living world itself as a reality of enormous promise. No one has developed this argument with more intellectual rigor and eloquence than John Haught. Recall how *On the Origin of Species* lays out a compelling narrative of the way that life over billions of years has felt its way forward toward greater complexity, beauty, and sentience. While there was no clear blueprint, writes Haught, the human capacity for discerning patterns can see in retrospect that over the long expanse of ages there has been a sort of directionality to the story: "it is undeniable that matter has gradually become alive, and within the last 200,000 years it has even begun to think and pray."[37] Even before the appearance of humans, life displayed an anticipatory quality, a dynamic reaching forward toward more sophisticated organization and function. It is no accident that this same cosmic dynamism now finds a new blossoming in human beings with our deep restlessness, yearning desires, sense of adventure, and longing for fulfillment. The story of cosmic and biological evolution makes it apparent that from the beginning the universe was seeded with promise, pregnant with surprise. This promissory character of the natural world, envisions Haught, is due to the inexhaustible vitality of God who created it: "From a Christian theological point of view, life and evolution are the universe's response to the presence and promise of divine persuasive love."[38] And the story is not over yet. Since the totality of nature and its long history are God's creation, and not our own, Haught points out, we can assume that it has levels of meaning and value that we humans may never fully grasp. Before humans beings appeared, evolution had brought forth countless diverse creatures, most of them having little or nothing to do with our own existence yet loved by God. Who knows what further significant developments await emergence in the future? In view of the still unfinished creation of life, we have the responsibility to leave ample room for more incalculable outcomes:

> Even if these outcomes have little relevance to our own lives and interests at the present moment, a robust creation faith demands that we rejoice in the prospect that other natural beings have a meaning and value to their Creator that may be quite hidden from our human powers of discernment. This universe, it bears constant repeating, is God's creation and not our own. It has taken billions of years for nature to attain the ecological richness that existed prior to our appearance. So when in our own time we allow pollution, resource exhaustion, and the annual extinction of thousands of species to fray the delicate tissue of life, we are surely aborting the hidden potential for a larger and wider-than-human future creativity that still lurks in the folds of the earth's complex ecosystems.[39]

The evolving world bears in its present perishable glory the possibility of an historical flowering. Significant transformation is still going on. Allowing the embryonic future to perish now at the hands of our own carelessness and selfishness is not only a violation of nature's sacramental being, but also a turning away from the promise that lies embedded in all of creation.[40] On an adventurous journey toward unimaginable fulfillment, the promise already glimpsed in nature's beauty and vitality needs to be treasured and safeguarded for the sake of its future in God.

Many human beings in our day, of course, are intensely concerned about the fate of the natural world not only out of self-interest, fair enough in itself, but also because of the world's beauty and intrinsic worth. Those who speak of faith in God have every reason to join this cadre of carers. A major element of the good news to which Christians bear witness is that they themselves are profoundly loved by the Creator of heaven and earth, a God of steadfast mercy and kindness. Turning attention to the natural world they can see that it is cherished by the same inexhaustible love. In its continuous creation by the empowering presence of the Spirit, its redeeming solidarity in the flesh of the crucified and risen Jesus Christ, its origin and ultimate future in the faithful love of the Creator, and its sacramental and revelatory character in all concrete beauty, suffering, and surprise—from every theological angle the tree of life calls forth deep respect and responsible love. As its bare, natural, evolving self, it is worthy of this. At the same time, if the diversity of creatures is meant to show forth the goodness of God which cannot be well represented by one creature alone, as Aquinas saw, then extinction of species is rapidly erasing testimony to divine goodness in the world now and for the foreseeable future. This connection between Creator and species undergirds William French's judgment that the march of vast numbers of species toward extinction is theologically idolatrous, brought about by policies that place lesser goods and in particular the gods of money and comfort above the God of life. Bowing down to false gods we are letting loose the forces of nonbeing with unprecedented viciousness and magnitude.[41]

CONVERSION TO THE EARTH

Looked at in this light, the ongoing destruction of life on Earth by human action, intended or not, has the character of deep moral failure. To speak theologically, it is profoundly sinful. By acts of commission and omission we are perpetrating violence against life and its future. In so doing we are pulling contrary to the will of God, whose beloved creation this is and whose goodness is reflected in its diverse forms of living species. Ethicists have coined new words to name the sin: biocide, ecocide, geocide. Sacrilege and desecration are not too strong a designation. Speaking theologically, the Catholic bishops of the Philippines name the despoilation an insult to Christ: "the destruction of any part of creation, especially the extinction of species, defaces the image of Christ which is etched in creation."[42] Whatever the language, the religious judgment remains that the damage humans are wreaking on the earth is profoundly wrong.

In a message issued for the 1990 World Day of Peace, John Paul II flagged this dimension of the ecological crisis. "*The ecological crisis is a moral issue*,"[43] he declared, backing up this judgment with strong, descriptive phrases such as "dramatic threat of ecological breakdown," "plundering natural resources," 'increasing devastation of the world of nature,' "uncontrolled destruction of animal and plant life," "reckless exploitation," and "the profound sense that the earth is 'suffering.'" At root, the pope suggests, the problem stems from lack of respect for life. This is, one might say, the cardinal sin. Characteristic of Catholic teaching, papal analysis here forges a strong link between human life blighted by structures of poverty and the integrity of nature disrupted by those same structures forged by market prowess:

> Often, the interests of production prevail over concern for the dignity of workers, while economic interests take priority over the good of individuals and even entire peoples. In these cases, pollution or environmental destruction is the result of an unnatural and reductionist vision which at times leads to a genuine contempt for human beings. On another level, delicate ecological balances are upset by the uncontrolled destruction of animal and plant life or by a reckless exploitation of natural resources.[44]

Social injustice and ecological degradation are two sides of the same coin, lack of respect for life. Both evils precipitate out from policies and lifestyles that reward the greed and selfishness of some to the disadvantage of many others. Drawing on the emphasis on the value of human life characteristic of Catholic teaching, John Paul II articulates a vigorous new principle of moral behavior:

> Respect for life and for the dignity of the human person extends also to the rest of creation, which is called to join humanity in praising God. [45]

This norm marks out a stunning new ethical horizon. It implies that moral consideration must be given to species beyond the human, and moral standing to ecological systems as a whole. In terms of the moral good, we owe love and justice not only to humankind but also to otherkind. The moral responsibility associated with extending respect to the natural world thus calls into play the substantial tradition on right and wrong, virtue and sin, already so well developed in terms of the dignity of the human person, and invites its challenging application to this new set of lives.

In a deft interpretation of prayer that flat out uses the language of sin, Bartholomew, Ecumenical Patriarch of Orthodoxy, holds up a mirror to the face of those who pray for the good of creation. When we pray God for the preservation of the natural environment, he writes in a 2012 encyclical, we are essentially praying for "repentance for our sinfulness in destroying the world instead of working to preserve and sustain its ever-flourishing resources." On one level we are imploring God "to change the mindset of the powerful in the world, enlightening them not to destroy the planet's ecosystem for reasons of financial profit and ephemeral interest." At the same time, each one of us generates small ecological damage, wilfully or ignorantly. "Therefore, in praying for the natural environment, we are praying for personal repentance for our contribution – smaller or greater – to the disfigurement and destruction of creation."[46]

When it comes to living the Christian life, what pope, patriarch, and numerous other religious leaders are urgently preaching is the need for people to change their ways. The traditional term for such a change is conversion. In a broad sense conversion is a continuous characteristic of the life of faith, an ever-deepening fidelity in relationship with God. Quite specifically, as the New Testament term for conversion (Greek *metanoia*) indicates, conversion also means literally a turning, a change of direction, switching away from one path and swiveling toward another. Accounts of religious conversion through the centuries make clear how this turning results from an awakening, slowly or abruptly, to certain spiritual realities, a new awareness that occasions changes in lifestyle, thought patterns, and moral commitments. In our ecological age we know that the Creator Spirit's presence and activity in the natural world has issued in creative abundance, biological diversity, ecological interrelatedness, and manifold possibilities. We humans sin when by acts of commission, omission, or sheer indifference we disappear species, reduce biodiversity, break up integrated ecosystems, and cut off future possibilities. Facing these evils in a spirit of repentance, we need the grace to be converted to the patterns established by the Spirit in the giving of life itself. We need a deep spiritual conversion to the Earth. This involves several discrete turnings at once.

☙ Intellectually, it entails moving from an anthropocentric, mostly androcentric view of the world to a wider theocentric one that has room for other species to be included in the circle of what is religiously meaningful and valued. It means letting go of a philosophy shaped by hierarchical dualism that prizes spirit over matter in favor of one that also intensely values physical and bodily realities as God's good creation. Rather than setting up a contrastive either-or relation between God and the world, this intellectual turning grasps the presence of the Giver of life in, with, and under the ecological community of species. Moving from denial that allows us to slack off under the weight of ignorance, it opens our eyes to the global impact of our everyday actions.

☙ Emotionally, being converted to the Earth involves a turning from the delusion of the separated human self and the isolated human species to a felt affiliation with other beings who share in our common status as creatures of God. In the beautiful words of Albert Einstein, "Our task must be to free ourselves from this prison by widening our circle of compassion to embrace all living creatures and the whole of nature in its beauty."[47] With this turning comes an experiential grasp of how deeply humanity is embedded in the evolutionary processes of life on Earth. In the depths of our being we recover a capacity for subjective communion with the natural world, to the point where brother sun and sister moon, brother fire and sister water, brother wolf and little sister birds are more than poetic ways of speaking but felt truths, as with Francis of Assisi.

☙ Ethically, ecological conversion entails the view that in our day a moral universe limited to human persons is no longer adequate. We need to widen attention beyond humanity alone and recenter vigorous moral consideration on the whole community of life. Recognizing that we are kin, we behave not just with utilitarian intent, though that is legitimate within limits, but with intent to preserve and protect creation which has its own intrinsic value. As Larry Rasmussen argued in his prize-winning *Earth Community, Earth Ethics*, ecological degradation is not just one more issue to be addressed along with the misery of racism, poverty, domestic violence, and other human ills. It embraces all these and more, insofar as our ecologically destructive actions are depleting and degrading the very conditions that make human life possible at all, to say nothing of jeopardizing the rest of life in fundamental and unprecedented ways: "one particularly powerful and errant species is overwhelming" the earth.[48] Coming to terms with this new wild fact requires a responsible ethical stance in which we learn to do with less in view of the good of the whole. Healing our moral paralysis, conversion opens ways for reciprocity rather than rapaciousness to mark our relationship with the earth.

In sum, ecological conversion means falling in love with the Earth as an inherently valuable, living community in which we participate, and bending every effort to be creatively faithful to its well-being, in tune with the living God who brought it into being and cherishes it with unconditional love. This turning is not done to the exclusion of other human beings, especially those poor and marginalized, but in view of their flourishing which is intertwined with ecological health on all levels. With trenchant insight Denis Edwards has written that, "When human beings first emerged in evolutionary history with their capacity for self-reflection and freedom, they emerged into a world of grace."[49] The living God whose compassion is over all creation was already present, embracing them with divine love. Throughout the vagaries of history that same divine presence has never deserted the natural tree of life nor the evolved human species on one of its branches. Our wretched sinfulness is continuously matched by forgiving grace that calls to repentance. Being converted to the Earth and its myriad inhabitants at this time of their distress is a moral imperative that transforms us toward greatheartedeness, in resonance with the Love who made and empowers it all.

10

THE COMMUNITY OF CREATION

> Protect creation ... protect all creation, the beauty of the
> created world ... respect each of God's creatures and respect the
> environment in which we live ... care for creation and for our
> brothers and sisters ... protect the whole of creation, protect
> each person, especially the poorest ... Let us protect with love
> all that God has given us!
>
> Pope Francis, Inaugural Mass, March 19, 2013

"WE ARE ALL CREATURES"

This book has been making the case in dialogue with Darwin's
theory of evolution that loving life on Earth, far from being
foreign to the living tradition of Christianity, is actually supported
by its core cherished beliefs about God revealed in scripture and
condensed in the creed. Despite the resources for ecological
conversion that Christian faith carries, however, vocal critics have
censured this religion for abetting the ecological crisis rather than
easing it.[1] And indeed, with some spectacular exceptions, it is
mostly true that the Christian churches both in their institutions
and members do not face this crisis with the energy they pour out
on other matters. Despite good official statements, committed
personnel, parishes going green, celebrations of Earth Day Sunday,
and some excellent voluntary work on local and international

levels, the plight of the natural world is not high on the agenda of the majority of Christian believers. Conversion to the earth is of secondary importance, if it is considered at all. It is as if Earth were undergoing its agony in the garden, about to be crucified, and we, the disciples of Jesus, are curled up fast asleep.

A key formidable obstacle to a change of heart, in the view of numerous theologians who have begun to grapple with this issue, is the tradition's way of envisioning human beings as a species set apart to rule the natural world. Gleaned from the first chapter of Genesis where the mandate to "have dominion" is given to the first human couple, the predominant human role has been seen as that of command and control. Due to their innate superiority humans have the right to master the natural world, which in turn is created to serve human purposes. It is true that the notion of dominion can be interpreted beneficently as a call to stewardship, a responsible vocation to protect and care for creation. As the role of dominion has seeped beyond the churches into cultural practice, however, it has been taken mainly as a right to dominate nature, with dire results. Either way, dominion pictures human beings at the apex of the pyramid of living creatures with rights over otherkind. This self-understanding has seeped into the depths of the Christian approach to nature, accounting for the tenacious anthropocentrism that attends most theologies.

In a felicitous development, biblical scholars in our day have discovered that the paradigm of dominion is not the only nor even the main view proposed by the Bible. More common is the paradigm of the community of creation, based on the understanding that humans and other living beings, for all their differences, form one community woven together by the common thread of having been created by God. This is not to say that the Bible is unambiguously "green" at first reading. The various books that comprise inspired scripture concern themselves mainly with the interaction between God and the human world, giving the whole an undoubted anthropocentric focus. Those who find in the text today an unalloyed ecological sensibility seem just as simplistic as critics who dismiss scripture out of hand because of the mandate to dominion. The Bible is a complex set of works, written over centuries in different genres with various intents. The crucial factor is hermeneutical, how it is interpreted. As the history of interpretation makes clear, the presuppositions that one brings to reading the text and the methods one uses will unlock different shades of meaning. At times a new question will unleash new insight. The 19th-century question about the moral rightness of slavery, for example, led to the conviction that it was contrary to God's intent, despite explicit biblical texts that accept and lay down rules for that evil institution. In a similar way, the ecological crisis of our era raises the theological question of the religious meaning of the natural world, and the accompanying ethical question of how people should rightly relate to the rest of creation. This new issue opens a fresh interpretive possibility. It allows for a reading of the Bible that spies an option largely unnoticed in the tradition, one having good expansive consequences for the human spirit and human behavior toward the rest of the evolutionary world. Working in this vein, scholarly efforts have concluded that dominion is only a partial model that does not exhaust the range of biblical possibilities. When interpreted as a whole, the Bible situates the function of dominion within a broader vision of a community of all living creatures centered on God.

Asking the beasts, birds, plants, and fish one last time this chapter examines these two paradigms, evaluating their ability to guide ecologically sound ways to envision and enact the human relationship with all other species on the earth.

THE DOMINION PARADIGM

The founding text for the dominion model is Genesis 1.28. On the sixth day after creating male and female human beings in the divine image and likeness, God blesses them saying, "fill the earth and subdue it; and have dominion over the fish of the sea and over the birds of the air and over every living thing that moves upon the earth" (1.28). The whole gorgeous array of creation precedes this text. By the divine word God has created the heavens and the earth with all their moving, flying, swimming inhabitants and pronounced them good. Now humans are added to the assembly, and told to exercise some kind of authority over the rest. What does this mean?

At the time this text was composed the natural world was seen as a wild place, at times threatening to human beings, unlike the reverse situation today when human activity poses a danger to nature's survival. There were lions in Israel! Their predation and that of other wild animals, coupled with stony soil that resisted cultivation, the sea that could erupt with life-threatening storms, and the vagaries of weather that could ruin needed crops placed humans in a precarious position vis-à-vis forces of nature. Some scholars surmise that in its historical context the dominion text gives legitimacy to the human need to secure a protected place in the midst of a powerful natural world that intimidated them and that they could not completely control.

A second line of interpretation takes its cue from a custom of the ancient royal court. Unable to be present throughout an extensive territory, a king would appoint an official to oversee the region in his name. Such an official would represent the king. He would be said to have "dominion" over that part of the kingdom, charged with carrying out the wishes of the ruler he stands in for. In this light, the Genesis mandate to have dominion clearly does not give human beings permission to *dominate* the natural world. God has just created all living things, blessed them and their fertility, and pronounced them all good. Having dominion in the royal sense means that humans are to be God's representatives, carrying out the divine will that other creatures should flourish. At this point in the narrative people do not even eat animals, since God gives them only plants and fruits for food (1.29). Far from a warrant to exploit, the Genesis mandate to have dominion gives humans "a delegated participation in God's caring rule over his creatures."[2] A later story in Genesis underscores the kind of responsibility entailed in the royal meaning of dominion. Prior to the great flood Noah is told to bring two of every kind of animal into the ark "to keep them alive with you" (Gen. 6.19). This he does, including in the boat's notable menagerie not only domestic animals but also wild beasts of no earthly use to humans. "Noah's gathering of the animals to save the species makes clear at last what having dominion over the animals means," writes Richard Clifford with brilliant insight: "seeing to their survival."[3]

It is important to note that the second creation story which follows in Genesis 2, far from using the model of dominion, places humans and the natural world in a quite different pattern of relationship. Recall how in this narrative a certain Hebrew word play emphasizes the earthy kinship between humans and other animals, both being made of the same stuff. In colorful verse the Creator gets the divine hands dirty by sculpting the earth creature (*'adam*) from the dust of the ground (*'adamah*), and breathing "the breath of life" up its nostrils. The earth creature is placed in a garden, with the mandate to "to till and keep it," that is, to cultivate and care for it. This earth creature seems lonely, and the Creator thinks to make a partner for mutual help and company. So out of the dust of the ground (the same 'adamah) God forms every animal of the field and every bird of the air. These living beings also have "the breath of life" in their nostrils, as later noted by Gen. 7.22, obviously breathed there by the One who gives life. When the Creator places the earth creature into a deep sleep, removes a rib, and fashions it into a female, the human couple, now sexually differentiated as man and woman, share the same bones and the same flesh with each other, but also with the animals and birds, all made from the same dust of the ground and thrillingly alive with divine breath of life. This account underscores the earthy solidarity women and men have with each other and with the rest of creation. Far from being taken out of the natural world and placed over it, they are made of the same stuff, immersed in a web of reciprocal relations with the land and other creatures, and charged to reverence and serve, to carefully use and protect them. Already the second creation story enfolds the dominion mandate of the first chapter of Genesis into a more mutual pattern of relationship.

The paradigm of dominion is explicitly reiterated in Psalm 8: "You have given them dominion over the works of your hands; you have put all things under their feet" (v. 6). The latter phrase, used elsewhere of a king's victory over enemies, clearly indicates vigorous conquest by a superior force. As with "subdue" in Genesis 1.28, this text is difficult to interpret in an ecologically beneficial sense. Critical analysis shows that both Genesis 1 and Psalm 8 were composed by priestly authors connected with the Jerusalem temple. Thus they reflect a primarily hierarchical view of the world, whether from a priestly or royal perspective.[4] Even when interpreted in a beneficent sense, the idea of dominion posed within such a hierarchical framework places human beings, or at least an elite male cadre of them, in a position outside of and superior to other species, which are meant for their service.

Perhaps because the dominion text in Genesis 1 stands at the beginning of the Bible and is embedded in such a majestic, memorable narrative, its view of the human-world relationship has long held sway in the Christian interpretive tradition. The focus on this model by a long line of interpreters has resulted in an imagination that simply assumes the supremacy of human beings over the rest of nature, with the corresponding right to use its resources to their own advantage, however carefully. That these creatures might have their own reasons for existing apart from human use does not enter the picture.

The effective history of the dominion paradigm through the centuries shows its ambiguity, insofar as it is wide open to readings that promote human self-interest at nature's expense. It has certainly been used, explicitly and implicitly, as an ideological justification for exploitative practices. Such a reading became especially ascendant in the modern era, when new methods of scientific investigation coupled with industrial development, new technologies, and global trade for profit promoted the idea that men (I use the word advisedly) had the right to master the natural world. The fact that there might be a cost to nature itself was passed over in silence. For the most part the Christian churches went along with this view because of their interest in promoting human betterment. Committed to charity, they focused on alleviating human needs without attending to the whole picture of the rest of life. Thus standard teaching has held that we humans are the superior species; plants and animals are made to serve our purpose; we have the right to rule and control them, even though we must do so prudently. No harm was necessarily intended to be foisted on the plants and animals. However, in the rough and tumble of the project of mastering nature, the idea of dominion became an unspoken warrant for destructive environmental practices. In the absence of a strong countervailing interpretation in the churches, human greed and pride almost invariably tipped dominion over toward domination. "The modern culture of materialistic excess has developed in the context of a notion of dominion as an unrestricted right of masters and owners to exploit all the resources of creation,"[5] critiques Richard Bauckham. Slipping its biblical bonds, the notion of dominion has supported rampant use and abuse of the earth.

The idea of stewardship as extensively developed by Christian evangelical thinkers today goes a long way toward restoring the balance upset by grossly egotistic interpretations of dominion.[6] A steward is a person who manages another's property or financial affairs, one who administers material wealth as the agent of another. The core of theological stewardship is the belief that the Earth and all of its resources belong ultimately to God. With overwhelming generosity God entrusts these good things to human beings, gifting us with their use. The first response is awesome gratitude. Then comes the realization that since ultimately we do not own these good things, the human vocation is to take care of them in the name of their Owner. Human beings are stewards, charged with maintaining the good condition of Earth's resources, respecting their limits, and using them wisely with a view to sustaining them into the future. This is called, with beautiful alliteration, "creation care." Human beings are responsible before God to shepherd the ecological treasure, "to till and keep it" (Gen. 2.15).

The stewardship interpretation of the mandate to have dominion honors the singularity that the human species undoubtedly is while firmly connecting our powers with a moral responsibility to act for the well-being of other species. It preaches well. In a fine way it envisions human beings as shepherds of creation, entrusted with its vitality, called to its care and protection. This provides a valuable framework for good ethical practice—and would that this were more widespread.

Problems with this model, however, have also become manifest. Primarily, it omits from view the clear interdependence of the human species with the rest of life on Earth. Even at its best, it envisions human beings independent from the rest of creation and external to its functioning. Lacking a deep ecological sensibility, it establishes a vertical top-down relationship, giving human beings responsible mastery over other creatures but not roles alongside them or open to their giving. The one-sidedness of the relationship makes the natural world a passive recipient of our management. Not incidentally, this pattern also obscures the Creator's relation with the natural world prior to and apart from human mediation.[7] Such critique of the notion of stewardship in no way intends to deny simplistically that the human species does indeed exercise power over other species. The ecological crisis itself gives evidence that our cultural prowess is overwhelming the very ability of other creatures to survive and reproduce. The criticism does suggest, however, that making human authority over nature the central pattern does not go the distance at this time of monumental dying off.

Among ecological theologians there is serious doubt whether the relationship of dominion on its own, even if redefined, is sufficient to change human sensibility and its consequent behavior in our day. We are sinners, after all, and being in charge offers an ever-present temptation to self-aggrandizement. The strong hubris entailed in the effective history of this paradigm needs to be remedied by a different conceptuality of the human place in the world, religiously speaking. Such an alternative presents itself in the biblical view of the community of creation. Widespread in prophets, psalms, and wisdom writings, this paradigm positions humans not above but within the living world which has its own relationship to God accompanied by a divinely-given mandate to thrive. Refashioning the idea of human relation to the natural world along these lines not only provides a context for a non-negotiably responsible retrieval of dominion but also opens the imagination to multiple avenues of reciprocal interaction between human beings and other species. Broadening the terms of our own identity in light of the reality of others, we end up seeing, thinking, and acting differently.

THE COMMUNITY OF CREATION PARADIGM

If evolutionary science has established any great insight it is that all life on this planet forms one community. Historically, all life results from the same biological process; genetically, living beings share elements of the same basic code; functionally, species interact without ceasing. Human beings belong to this community and need other species profoundly, in some ways more than other species need them. Take trees as an example. To stay alive trees take in carbon dioxide, synthesize it in the presence of sunlight, and give off oxygen as a result. Earth's atmosphere is rich in oxygen thanks to the lives of green plants, trees tallest among them. Human beings breathe in this oxygen and exhale carbon dioxide as a waste product. Which species is more needy of the other? In a thought experiment, remove humans from the earth. Trees would survive in fine fashion, as they did before humans arrived and started to cut them down. Now imagine trees removed from the planet. Humans would have an increasingly hard time surviving, with growing amounts of carbon dioxide in the atmosphere and less oxygen to breathe. The point is, human beings are not simply rulers of the life-world but dependent upon it at the most fundamental level.

The biblical vision of the community of creation offers an analogous view of the interdependence of life for more ancient religious reasons. In its origin, history, and goal, the whole world with all its members is ultimately grounded in the creative, redeeming God of love. Neither plants, animals, nor human beings, neither land, sea, nor air, neither sun, moon, nor stars would exist apart from the life-giving, loving power of the Creator. When parsed to its most basic element, the relational pattern of the community of creation is founded on the belief that all beings are in fact creatures, sustained in life by the Creator of all that is. To reiterate: at the core of their identity, humankind and otherkind share this same fundamental status of being finite creatures. As such, *human beings and other species have more in common than what separates them.*

In this profoundly theocentric view, human beings participate with others in an interdependent world fundamentally oriented to God. We are situated within, not over, the magnificent circle of life, whose center and encompassing horizon is the generous God of life. This is a kinship group of hugely diverse members whose mutual relationships are enormously rich and complex. In varied interactions each member gives and receives, being significant for one another in different ways but all grounded in absolute, universal reliance on the living God for the very breath of life. Within this guild of life the distinctive capacities of human beings are part of the picture and can be exercised without lifting our species out of creation, "as though we were demi-gods set over it."[8] If humans are defined first of all as fellow-creatures, dominion beneficently understood becomes a role within the larger sphere of community relationships, which are reciprocal rather than one-way.

Repositioning the human species within the community of creation centered on the living God and reconceiving our identity primarily along the lines of kinship rather than rule opens a promising new avenue for religious self-understanding and sound practice. Rather than setting up a top-down structure of relationship, theology done in this framework makes its first word that "of our connection to, relationship with, and solidarity alongside others of God's creatures, rather than of differentiation from them, which has been the more common starting point (and frequently the ending point too)."[9] While embracing the best of stewardship theology and its ethical behavior, this model's different imaginative framework unleashes aesthetic, emotional, and ethical responses that express ecological sensibility at a fundamental level. Note that the relationship envisioned here does not encourage communion by the ploy of blurring the lines between species, as if *Homo sapiens* were not a singularity. Rather, it allows each species to stand in its own difference, but encompassed by a wider whole that affects their interrelation.

To take the measure of this paradigm, consider key biblical texts that present a strong sense of the community of creation and offer other mandates besides dominion. Allowing these writings to fertilize our imagination will suggest new patterns of theological anthropology rife with potential for critical life-enhancing spirituality and practice.

"WHERE WERE YOU ... ?"

No biblical book presents the community of creation more firmly and eloquently than the book of Job, which contains the longest piece of writing on the natural world in the Bible. Its theological vision offers a strong antidote to the human arrogance that has flowed in the modern era from the view of dominion as domination. As the ancient folk tale unfolds, Job is suffering greatly. Mouthing the standard conviction of their culture, his three friends argue he must have sinned greatly to deserve this punishment. In a debate that grows increasingly acrimonious, Job maintains his innocence. Flinging anguished accusations against divine justice, he brings a lawsuit, challenging God to appear in court to defend the way the world is ordered.

> Then the Lord answered Job out of the whirlwind ...
>
> (Job 38.1)

The answer is unexpected. In gorgeous poetic language over the course of four chapters (38–41, which readers are encouraged to peruse for themselves), the text paints a picture of God's activity in creation, emphasizing that the human role in the life of other species is next to nothing. The voice from the whirlwind sets the theme with a daunting question:

> Where were you when I laid the foundation of the Earth?
>
> (38.4)

This query repeats over and over again, putting Job and through him all human beings in their proper place vis-à-vis the Creator and other created beings who are beyond human control. Where were you when the Earth was measured out, when the stars began to sing together, when the sea was placed within boundaries and its proud waves given limits? Have you commanded the light to rise at dawn? the snow and rain to fall even where no one lives? the thunder and lightning to play? Orion and the other constellations to run their courses across the sky?

Once the physical world is laid out, the questions from the whirlwind turn to the behavior of animals who for the most part are wild and free, living out their lives without serving human purpose. Who provides prey for the lion, hunting food for her young who lie waiting in their den? Who gives prey to the raven, whose young ones are crying out with hunger? Do you know when the mountain goats crouch and give birth, their young then growing strong and roaming away? Have you given the wild ass its freedom? It roams the steppe for pasture, scorning the distant city and the shouts of a human driver. Is the wild ox willing to serve you, to be tied up at night and plough your fields by day? Look how the ostrich flies, laughing at riders on horseback. Do you give the majestic war horse its might? Is it by your wisdom that the hawk soars or by your command that the eagle mounts up, spying their prey from afar, with their young sucking up blood from the slain?

Two further questions ask whether Job can contend with magnificent, fearsome beasts. There is no scholarly consensus about the exact identity of these creatures, which evoke mythical monsters. The first may well be modeled on the hippopotamus: "Look at behemoth, which I made just as I made you" (40.15). This creature is amazingly strong, with powerful belly muscles, bones like bronze, and limbs like bars of iron. It wallows under the lotus plants, in the reeds and marsh, surrounded by the willows of the wadi; even if the river is turbulent, rushing against its face, it is not frightened. Human wiles cannot capture it: "Can one take it with hooks or pierce its nose with a snare?" (40.24). The second animal, leviathan, seems to be modeled on a large, ferocious crocodile, with neck and limbs of mighty strength, a double coat of scales that cannot be penetrated, a terrible set of teeth, and gleaming eyes. Far from being amenable to servitude to humans—"Will you play with it as with a bird, or will you put it on a leash for your girls?" (41.5)—it laughs at attacks from spears, arrows, clubs. There seems to be a bit of laughter coming from the whirlwind when the voice challenges Job, "Lay hands on it; think of the battle; you will not do it again!" (41.8). Beyond their own immediate reality, both creatures appear to symbolize forces of chaos that threaten the order of the world. Job can no more control them than he can the lion or wild ox.

As centuries of profound commentary on this book have made clear, the divinely sketched panorama of the created world from its beginning up to its current ordering leaves us with three results. In terms of the book's presenting problem, the suffering of a good person, there is in the end no direct answer in the sense of a rational explanation; the mystery of evil remains unfathomable. At the same time, one reason traditionally adduced for such suffering, namely, that it is sent from God as a punishment for sin, is clearly rejected. The Lord upbraids the three friends on this point, "for you have not spoken of me what is right, as my servant Job has done" (42.7). While respecting mystery and disallowing chastisement, this biblical book does make one positive move regarding suffering. It places Job's pain in the context of God's nearness in cosmic creation ... and he is filled with wonder. Stunned by encounter with the immensity, beauty, and intricate order of things, his stance is reoriented: "I had heard of you by the hearing of the ear, but now my eye sees you" (42.5). Shifting from an anthropocentric to a cosmocentric perspective, he now knows a different God, bigger than the tit-for-tat ruler both he and his friends had envisioned. "He is taken out of himself and given a broader vision of the universe and God's ways with it. What brings home to him the incalculable wisdom and power of God is the *otherness* of the cosmos, precisely that it is not a human world."[10] This expands Job's horizon to the point where he deeply grasps that God's love does not act according to rules of retribution which a penal view of history insists upon, but like all true love operates freely in a world of grace that completely enfolds and permeates him, even in pain. With new clarity of vision, his story moves toward healing and peace.[11]

Pursuing the problematic issue of the suffering of the innocent, the author of Job is obviously not interested in our current ecological question about the right relation between human beings and the rest of creation. The point to be noted is that in the course of its own reflection, this biblical book assumes, presents, and builds its argument on a vision of relationship that places humankind in a remarkably different position from the dominion text of Genesis 1. While including similar elements from the physical world and the world of life, the creation narrative unspooled from the whirlwind sees human beings within a different framework. The biggest difference is the absence of the mandate to have dominion. Instead of being placed at the apex of creation, Job is led to see divine activity in the awesome, independent working of the natural world over which he has no mastery, not only technologically but also theologically: "Where were you ...?" The whirlwind's vision of creation's grandeur makes a religious point, namely, that the human place in the scheme of things is not first of all one of supremacy. We are not the center of everything. It is not all about us. Rather, we belong in the first instance as fellow creatures alongside God's other creatures. The wild animals mentioned, as Clare Palmer writes, "are completely independent of humanity: the hawk, the mountain goats, the wild ox, the leviathan; they are not made for humanity, not made to be human's companions, nor even made with humans in mind. They live their own lives."[12] Each animal has a unique value, even the most fearsome. Their free wild spirits defy human domination and can survive without us. Encounter with the otherness of their wildness can even evocatively mediate "the qualitatively different otherness of God."[13] At the same time, humans are an integral part of this creation cared for in all its integrity, wisdom, and beauty by the Creator. It should not be overlooked that this non-anthropocentric biblical paradigm of the community of creation is tremendously confident of its own truth, ascribing its articulation to God's own voice.

Ecologically-aware scholars today find in the community of creation so superbly presented in Job a bracing summons to practice the virtues of humility and joy. Directing our urgent quest for right relationship, this paradigm introduces a powerful dose of humility to human beings. Its talk of multitudes of creatures who by divine design live freely, being subjects of their own lives rather than objects for human use, punctures human obsession with ruling over others. "We have lived for so long with this picture of ourselves, as subjects inhabiting a world that is our object and resource, that it is difficult to imagine it might not be true," Sallie McFague compassionately admits.[14] However, the repeated questions from the whirlwind—where were you? do you know? can you make it happen? can you provide?—urge a different view and can be read as an antidote to the pride and consequent disregard of other species that has found justifi-cation in the Genesis dominion text (though such an interpretation is not necessarily the only one). Humans take their place as creatures among other beloved creatures in whom the living God is independently interested.

At the same time, the community of creation image is bursting with God's delight in the flourishing of life in the natural world, a joy which humans are called to share. In the book of Job this is not said in so many words, directly. But the voice from the whirlwind's close descriptions of animals' idiosyncratic behaviors and habitats, the colorful abundance of word pictures, the sheer poetic power of these verses brings a subtle realization of divine enjoyment to the fore. The skill of the hunting lion, the freedom of the wild ass, the soaring flight of the eagle, even the apparent stupidity of the ostrich who lays her eggs where feet can crush them: the divine voice lingers lovingly over each one with what sounds like pleasure and pride. By sharing divine admiration of these creatures with similar gladness, Job, and with him human readers of the text, enter into deeper relationship with the Creator's joy in the world.

As Bill McKibben notes, these responses evoked by the whirlwind are crucially needed at our time of ecological distress. Humility before the cosmos and joy in the workings of nature may seem contradictory. The former can be aridly negative, the latter irresponsibly gleeful. Yet taken together, "They are reinforcing, powerful – powerful enough, perhaps to start changing some of the deep-seated behaviors that are driving our environmental destruction, our galloping poverty, our cultural despair."[15] Humbled and delighted by the other life around us, we can grow to know ourselves as members of the community of creation and step up to protect our kin.

CREATION'S PRAISE AND LAMENT

Psalms

The psalms, prayers of ancient Israel continuously used in Christian liturgy, are another source of the biblical vision that positions human beings within, not above, the rest of creation. The great creation Psalm 104 is the most telling in this regard. Filled with ecological details like the book of Job, it does not pose challenging "where were you?" questions. Instead, it uses the diversity of species, including humans, to bless the extravagant greatness of God who creates and provides. From first to last this psalm bears a pervasive sense of the abundance and fertility of the world as God's generous gift to all living creatures, a sense strengthened by the way it punctuates lyrical descriptions of nature with volleys of praise.

First mentioned are the sky, the clouds, and the wind, along with fire and flame; the firm earth comes next, with its bounded sea, rain waters, and rivers. Like little nature photographs, textual vignettes display the vital connections between the land, water, vegetation, and animals:

> You make springs gush forth in the valleys;
>> they flow between the hills, giving drink to every wild animal;
>> the wild asses quench their thirst.
> By the streams the birds of the air have their habitation;
>> they sing among the branches."
>
> (vv. 10-11)

Human beings make their appearance in two verses near the middle of this psalm. Like other living beings they are dependent upon the land and water for their sustenance. Like others, their own ecological niche is described with wonder and gratitude. God makes grass to grow for their cattle and plants to grow for their own food; God causes the earth to yield wine to gladden their heart, oil to make their face shine, and bread to make them strong (vv. 14-15). With no break the psalm continues with praise of God for well-watered trees, great cedars, nesting birds, storks in the fir trees, wild goats in the high mountains, and coneys (a kind of rabbit) taking refuge in the rocks. The psalm notes that the natural rhythm of day and night is also a gift from God, who made the moon to mark the seasons and the sun to rise and set, which in turn triggers creatures' behavior, humans included:

> You make darkness, and it is night,
>> when all the animals of the forest come creeping out.
> The young lions roar for their prey,
>> seeking their food from God.
> When the sun rises, they withdraw
>> and lie down in their dens.
> People go out to their work
>> and to their labor until the evening.
> O Lord, how manifold are your works!
> In wisdom you have made them all;
>> the earth is full of your creatures.
>
> (vv. 20-4)

Next the psalm visits the great, wide sea, teeming with innumerable creatures large and small, traversed by ships, and the home of leviathan whom, delightfully, the Creator formed to sport in it. All of these wondrous beings are in God's care. They die and return to dust when God takes away their breath, but are created "when you send forth your spirit " (v. 30). The great panorama of creation leads to final affirmations of joy: I will sing and rejoice in my God! Alleluia!

In both structure and content, this psalm betrays no trace of a mandate for human dominion. It is a theocentric depiction of the world which positions human beings with their distinctive needs and blessings within a wider world which enjoys its own direct relation to the Creator. Humans are part of this wonderfully diverse creation. Like Job, this psalm stands as a counterweight to mastery carried out on the assumption that humans have a right to rule other species. Here, however, "there is no trace of human supremacy over the creatures in general. The impression is rather of fellow creatureliness."[16] Perhaps the strongest recognition that humans are exceptional comes at the end of the psalm, where it hits a jarring note with the wish that sinners be removed and the wicked be no more (v. 35). In context, this verse seems to be noting that human beings are the ones who disrupt the harmony of creation, ruining what is so beautiful by nature. We are the ones who sin, and would that we would stop. By contrast, as the final praise makes clear, the world as a whole is the work of God's hands, who made it and generously provides for its good. Its value lies not in human mastery nor simply in the benefits it can render our species. Praise rings out because the whole creation reflects divine glory and it is precious in God's eyes.

The biblical vision of the community of creation gains yet another dimension in the psalms that join human prayer to that of other creatures praising God. Besides describing the world's variety and blessing the Creator for its abundance, as the psalmist of 104 does, these psalms convey the sense that the creatures so described are themselves offering praise to the glory of God. The standout in this regard is the exuberant Psalm 148. Framed at the beginning and end by the joyful shout "Alleluia," the psalm works through an extensive array of beings extolling the Creator who gives them life. All angels, sun and moon, all shining stars, rain water in the sky, all sea creatures, fire and hail, snow and frost, stormy winds, mountains and all hills, fruit trees and all cedars, wild animals and all cattle, creeping things and flying birds, kings and all peoples, young men and women, old and young together: praise God's holy name. Why? Because God "commanded and they were created" (v. 5). All exist as the fruit of the powerful good will of the Giver of life whose name is exalted beyond heaven and earth.

Note that the order in which this psalm names different creatures is not anthropocentric, proceeding from lowest to highest with humans placed at the apex. While humans do appear last, angels, presumably a higher order of beings, come first; living sea creatures precede inorganic snow; creeping things follow more complex forms of mammals. Rather than the pattern of dominion which climaxes the appearance of humans in Genesis 1, and different from the hierarchical chain of being found in medieval thought, here is an interwoven assembly of everything from sky, sea, and land, each one part of a grateful community of creation praising God. Like members of a cosmic choir or a symphony orchestra, each makes a different sound, contributing in its own uniqueness to the grandeur of the created world.[17] Human voices swell the chorus in a distinctive way. But in this music of praise it is the living God who is being exalted, not any one creature over another.

There is some scholarly debate over how to interpret the very idea of the natural world being able to worship, having a voice which can be raised in praise or lament.[18] Is this not simply a poetic fancy? Presumably the act of worship requires rational consciousness at the very least, along with the will's desire to turn toward God. Truth be told, the image of stars, winds, trees, and animals praising the Giver of their life is indeed a metaphor, one drawn from the experience of humans at prayer. As such, however, it points to an ontological truth. By virtue of their being created, of being held in existence by the loving power of the Creator Spirit, all beings give glory to God simply by being themselves. In their very existence, their concrete quiddity, their working out their roles in an evolving universe, they extol the excellence of their Maker. Augustine saw this when he wrote that creatures praise God by existing and acting according to their natures:

> Let your mind roam through the whole creation; everywhere the created world will cry out to you: "God made me." ... Go round the heavens again and back to the earth, leave out nothing; on all sides everything cries out to you of its Author; nay, the very forms of created things are as it were the voices with which they praise their Creator.[19]

"A tree," comments Bauckham, "does not need to do anything specific in order to praise God; still less need it be conscious of anything. Simply by being and growing it praises God"[20] (so too, I would add, by dying naturally to provide nourishment as a nurse log for other creatures). In a word, "Creation's praise is not an extra, an addition to what it is, but the shining of its being, the overflowing significance it has in pointing to its Creator simply by being itself."[21] The very forms of plants and animals are their voice of praise.

The psalms' poetic metaphors have cognitive as well as aesthetic value.[22] Without the metaphor of worship highlighting nature's voice lifted in praise, our human minds could well overlook the reality of orientation-to-God embodied within the physical world. With it, other creatures take on a witnessing role, declaring to us the most profound truths. Individually, we are all fellow creatures of the same life-giving God. Together, we are all members of the community of life on Earth, engaged in complex interactions, sharing the world given as a gift by God. Attuned to nature's praise, human beings can glimpse a world expressing the glory of God. The voice of other creatures pierces through our distractions, and invites us to participate.

It is no accident that the theme of creation's praise of God makes a telling appearance at the three Masses traditionally celebrated in the Catholic liturgy on Christmas. Here the church's joyful prayer receives a certain fullness by joining with the exultation of Earth and its creatures. At midnight, for example, before the proclamation of the gospel of Jesus' birth in Bethlehem, the congregation prays:

> Let the heavens be glad and the earth rejoice;
>> let the sea roar and all that fills it;
>> let the field exult, and everything in it.
> Then shall all the trees of the forest sing for joy
>> Before the Lord; for he is coming ...

(Ps. 96.11-13)

Lest the grammatical structure of this and other psalms be misleading, it should be noted that the psalmist's invitation to other creatures to sing, be glad, rejoice, and shout, does not initiate their praise. Creation is already praising God with or without human attention. What the psalm does is bring this to our human awareness. Our voices swell the chorus in a distinctive way as we participate with other creatures in praise of God's goodness.

The positive notes sounded in many psalms do not address the incompleteness and struggle that pervade the ecological world. In that sense they are not only creation-oriented but eschatological in character, anticipating the final day when a renewed heaven and earth will rejoice in God's presence. The author of the last book of the Bible caught this inclusivity: "Then I heard every creature in heaven and on earth and under the earth and in the sea, and all that is in them, singing ..." (Rev. 5.13). On that day human voices will join this worshiping community composed of "every creature." In these psalms we do so already, tasting beforehand the wholeness of redeemed creation.

At this time of ecological catastrophe, praying with a sense of participation in creation's praise of God allows people to recover a healthy sense of their own human place in the world as created beings alongside our fellow creatures. Of central importance, when we join them in their own clapping and singing, we come to understand that their value for God is not based on their usefulness for us, an awareness with enormous ethical implications for how we exercise ecological responsibility. In ways unique to our cognitive singularity, we join their praise to our own more explicitly conscious thanks, becoming, in Abraham Heschel's telling phrase, the "cantors of the universe."[23] At the same time, the relation is mutual and redounds to our benefit. At a time when prayer does not come easily to postmodern humans, becoming aware of nature's praise may actually allow these other creatures to help us pray. By virtue of their being created, they are intrinsically related to their Creator. The more we attend to them, the more they can lift our hearts to God, borne on their praise.

Prophets

The biblical theme of the community of creation appears in different guise in the prophets, often laced with sorrow. The natural world and humans together share a common living space. When disaster strikes it affects them both. The prophets are most concerned with trouble that comes as a result of human wrongdoing. Then not only do people suffer, but the community of life does as well. As in the psalms, so too in the prophetic writings, creation cries out to God, but this time in lament:

There is no faithfulness or loyalty, and no knowledge of God in the land.
Swearing, lying, and murder, and stealing and adultery break out;
    bloodshed follows bloodshed.
Therefore the land mourns, and all who live in it languish;
together with the wild animals and the birds of the air, even the fish of
    the sea are perishing.

(Hos. 4.1-3)

Similar to creation's voice of praise, the image of the land mourning is a metaphor taken from human experiences of grief. In its poetic way it too tells an ontological truth, namely, that the natural world in its distress continues to be related to the God of loving kindness and faithfulness. The praise that creatures offer through their very being changes pitch through their devastation, becoming a wail of sorrow directed toward the Creator: the vineyard is ruined, "desolate, it mourns to me" (Jer. 12.11); "even the wild animals cry to you because the watercourses are dried up" (Joel 1.20). What the land and its inhabitants are bemoaning is ecological devastation caused by human action. Vegetation withers, springs of water disappear, animal life deteriorates, fruitful land becomes waste and void, because human beings flout the moral order. Our sinful behavior has destructive consequences not only for our own kind but for other creatures as well, bonded with us in one community of life.

This view runs through the prophetic writings from Isaiah, Jeremiah, and Ezechiel to Hosea, Joel, Amos, and Zephaniah. Flowing into the New Testament, it undergirds Paul's revelatory passage in Romans about the whole creation groaning and in travail, eagerly longing to be set free, its destiny intertwined with that of the human children of God (8.18-23). These vivid biblical depictions of creation's mourning, lamenting, groaning unto God in the midst of destruction could easily lead to despair. The prophets, however, announce a future redemption which will revivify the people and the natural world together:

The wilderness and the dry land shall be glad,
    the desert shall rejoice and blossom;
like the crocus it shall blossom abundantly,
    and rejoice with joy and singing.

(Isa. 35.1-2)

Prophets proclaim in hope that the mercy and steadfast love of God will establish justice in a disordered world, and this hope is announced with a vigor equal to their denunciation of human wrong-doing. The point to note here is that in both sinful and redemptive situations, human actions are embedded in the natural world and reverberate throughout the community of life. The ecological world, in turn, has its own independent relationship to the One who creates it, glorifying God in its very being and mourning to its Creator when some fellow creatures who are humans pollute and destroy. In view of the deep interrelation between the two in the community of creation, Bauckham makes a telling point about the praise of the creatures in Psalm 148. "For those who read or sing it with the desecration of God's world in mind, it is praise in defiance against evil and in hope of new creation."[24] We sing in praise with other species against the facts of their human-caused disruption and extinction. The negative contrast experience of the creatures' praise in face of ruination of the Earth impels us to action on behalf of the flourishing of all creation.

The biblical vision of the community of creation opens a life-enhancing avenue of relationship. Departing from a long history of interpretation, it scoops up the Genesis notion of dominion and places it back into the wider canon of scripture where it functions as one role within the mutual interactions of all beings as creatures in relation to the living God who creates and redeems. The community model brings forward at the most fundamental level our theological human identity as created, our biological embeddedness in the natural world, and our reciprocal inter-dependence with other species and the life-giving systems that support us all. Community becomes the radical context which puts the special role of human dominion, best understood as stewardship and responsible care, in its rightful place as one among many important exchanges we have with the natural world. This context likewise ensures that we would not be so likely to overlook the huge difference between human rule and God's rule: "Where were you ...?" (Job 38.4). Human responsibility is exercised within creation, in relation to other fellow creatures who are created as we are, and upon whom we depend for our own lives. As a role among creatures, it is shepherding for which we are ultimately responsible to God.

THE ECOLOGICAL VOCATION

Christian tradition has always interpreted the good we are called to do for other humans not first and foremost under the rubric of duty but as an expression of love, love of neighbor impelled by the love of God. Jesus' surprising parable of the good Samaritan reveals that the neighbor is the one who shows mercy to the assaulted, half-dead traveler by the side of the road; conversely, the neighbor is anyone in need (Lk. 10.25-37). If indeed "respect for life and for the dignity of the human person extends also to the rest of creation,"[25] as Pope John Paul II declared, then there is good warrant for extending the notion of neighbor beyond the human species to all other fellow creatures in the community of creation. In view of the world of life now under duress Brian Patrick riffs, "Who is our neighbor: the Samaritan? the outcast? the enemy? Yes, yes, of course. But it is also the whale, the dolphin, and the rain forest. Our neighbor is the entire universe. We must love it all as our self."[26]

Whether framed in these terms or not, numerous people around the globe are beginning to live the ecological vocation, caring for the living world as their neighbor. Consider these sketches in society at large. Parents spark love of nature in their children. Teachers instruct students with zest, using engaging curricula; colleges establish environmental majors. Creative artists write, paint, compose, make movies, dramatize. Architects design LEED-worthy buildings. Businesses incorporate environmental best practices. Investors aim their funds toward green companies. Farmers use organic rather than toxic materials on the land. Inventors figure how to use renewable energy from the sun, wind, and geothermal heat. Home owners grow native plants for wildlife and install solar panels. Schools, hospitals, and stores practice energy conservation. Communities plant neighborhood gardens. Scientists study to discover more of life's wonder and terrible wasting. Reporters and editors publish the results. Citizens recycle and give money to ecological organizations. Ecological activists work to protect habitats and migration routes. Ethicists articulate ecological virtues such as adaptability and frugality; the treatment of animals becomes a moral issue. Scholars in every field address the well-being of other species in their work. Lobbyists agitate for clean air and water. Legislators enact protective policies, executives enforce them, and judiciaries uphold their legitimacy. Nations forge international treaties. Lawyers defend the Earth in court.

Such an outpouring of practical acts is not yet a flood. But each one, which can already be well-documented, bespeaks a conversion to the Earth in an ecological sense as our *oikos*, our inhabited house, our only home in this vast universe. Each one reveals human beings in our day becoming beholders, looking out the window at living species which elicit and deserve long-term cherishing. Like latter day Noahs, these human beings are trying to save other creatures either directly or by protecting the environment in which they live.

As part of this movement, churches, synagogues, mosques, and temples, too, are taking ecological steps, now infused with religious sentiment. Congregations adopt best practices to conserve energy when heating, cooling, and cleaning their buildings, or watering and fertilizing their land. Leaders write ecologically inspiring letters and institute regional policies. Liturgists craft prayers, hymns, and rituals. Preachers preach from the book of nature. Pastors bless animals on the feast of St. Francis of Assisi. Theological education, educational ministries for children and adults, and social outreach programs highlight the value of creation and encourage care for its healing, even to the point of sacrifice.[27] Across a spectrum of people of faith traditional religious practices such as contemplation, asceticism, and prophetic action for justice return in an ecological key:

❦ In contemplation people look on the natural world with affection rather than with an arrogant, utilitarian stare. Using their senses and aided by scientific literacy, they learn to appreciate nature's astonishments and be alert to its harm. Religious contemplation rachets up what is at stake because it sees the world thus appreciated as God's handiwork, a place of encounter with the divine. The vivifying, subtly active presence of the Creator flashes out from the simplest natural phenomenon, the smallest seed. "For the little singing birds sing of God," writes John Calvin, while the animals acclaim, the mountains resound, the rivers throw glances, the grasses and flowers smile in praise.[28] Seeing that the bush still burns, we take off our shoes. Akin to prayer, contemplation deepens human connection with the world, enfolding other species into our love and passionate care.

Because we will not fight to save what we do not love, as Stephen Jay Gould has observed, such silent practice has strong long-term effects.

❧ The true purpose of asceticism has always been to make persons more fully alive to the movement of grace in their lives. It does so by sacrificial acts that remove what blocks sensitivity to the presence of the Spirit. Traditional forms of asceticism have come upon hard days due to their alliance with a philosophical dualism that prized spirit at the expense of matter. In light of ecological destruction, however, asceticism practiced with an eye for the good of other species acts with discipline precisely to protect physical life. A sensuous, earth-affirming asceticism leads people to live more simply not to make themselves suffer and not because they are anti-body, but to free themselves from enslavement to market practices that harm other living creatures.

❧ Christian social teaching about the common good underscores the need to change not just individual behaviors but social structures that create misery. If nature be the new poor, as Sallie McFague insists, then the passion to establish justice for poor and oppressed humans widens to include suffering ecosystems and other species under threat.[29] This may well entail action on behalf of justice that goes up against powerful vested interests. There are tough political and economic issues at stake, decisions about allowable business practices, budget expenditures, energy production, pollution controls, trade patterns, and the like where the well-being of the Earth is at stake. In the tradition of the biblical prophets and the spirit of Jesus, people band together and take critical public stances for the care, protection, restoration, and healing of Earth and its community of species, fully aware that they may elicit the classical reaction to the prophet and be despised.

One strong example: the American woman religious Dorothy Stang, SNDdeN, murdered in the Brazilian Amazon in February 2005. For over three decades Sister Dorothy was outspoken in her efforts on behalf of the rain forest and the indigenous communities that lived therein, opposing illegal logging and burning that were ruining their intertwined lives. Despite death threats from loggers and ranchers, she remained faithful to this ministry, often wearing a T-shirt that read, "The death of the forest is the end of our life." The cost was nothing less than the end of her own life as she was shot dead on a forest path, an open Bible her only defense.[30] The witness of her life joins that of others such as Chico Mendes, Vicente Cañas, and Wilson Penheiro, likewise killed for protecting the Amazon rain forest and its inhabitants. Other ecosystems also have human lovers who gave their lives.

The prophetic dimension of the ecological vocation still beckons the churches, for the most part. Financial investments, political interests, and social standing tend to tie their institutional public presence to the status quo, now so harmful to ecological flourishing. In the face of danger and loss, however, Christians personally and as church are called by the power of the Spirit to enter into solidarity with suffering creation and exercise responsibility for a new project of ecojustice. Participating in the compassion of the God of life, prophetic action takes risks to denounce wrongdoing and to announce in hope a more holistic vision of the community of life.

All of these actions and more express a deeply changing human consciousness. They signal that the time is past when humans could ignore the impact of their behavior on the rest of life and on the ecological systems that support all living beings. When brought to expression in the context of Christian faith, such practices bespeak a profound turning to the God of life. Inspired by the Spirit who pervades and sustains the community of creation, the human imagination grows to encompass "the other" and the human heart widens to love the neighbors who are uniquely themselves, not human. As I write, practices already underway are not accomplishing remotely enough. Too much has already been lost. But these are signs of hope. In the teeth of the ongoing ruination of the tree of life, faith in the living God can be a spur to action that makes a difference.

ONWARDS, FOR THE LOVE OF GOD

At the beginning of this book I invited you, good reader, to keep before your mind's eye some version of Darwin's entangled bank, some place where you noticed and responded with pleasure to the natural world. Interrogating this bank, we asked the beasts, the birds of the air, the plants of the earth and the fish of the sea what they could teach us. Their answer astounds. *Scientifically*, they have come into being, grown into their complex, beautiful forms, and fit into their diverse ecological niches through a powerful, unscripted evolutionary process that has lasted hundreds of millions of years. *Theologically*, they are the work of the Spirit of God who vivifies the community of creation from which we humans have also emerged. The ineffable holy mystery of Love creates, indwells, and empowers plants and animals, delights in their beautiful, wise, and funny ways and grieves their sufferings. In the unexpected Christian view, the living God even chooses to become part of their story in Jesus Christ, a member of the community of life on this planet whose death and resurrection pledges a hopeful future for all. *Ecologically*, the community of living creatures is now under terrible threat due to human action.

Revisit your variant of the entangled bank and see it through each of these filters. Is it not an inestimable treasure?

The story of this bank and the evolving world it represents is not over. Whither life? What will life on this planet be like in a century? In fourteen thousand generations? To a large extent human agency is now part of the evolutionary story. The future of the tree of life is now at the mercy of human decision and indecision. If ever there were a sign of the times to be interpreted theologically in light of the living God who creates and redeems, this is it. Impacted by the contours of the crisis, this book's dialogue between Darwin's view of evolution and Christian belief in the God of love has delivered us to a crossroads: the option for conversion to the Earth, or not. The option reaches into profound depths, for the call to be converted to compassionate care for other species is not in the first instance an ascetic or moral mandate, but an urgent invitation to be converted to God: to love in tune with God's abundant love so that all may have life.

The argument here has been that commitment to ecological wholeness in partnership with a more just social order is the vocation which best corresponds to God's own loving intent for our corner of creation. We all share the status of creaturehood; we are all kin in the evolving community of life now under siege; our vision must be one of flourishing for all. The immediate aim is to establish and protect healthy ecosystems where all creatures, including poor human beings and plants and animals being driven to extinction, can thrive. The long-term goal is a socially just and environmentally sustainable society in which the needs of all people are met and diverse species can prosper, onward to an evolutionary future that will still surprise.

A flourishing humanity on a thriving planet rich in species in an evolving universe, all together filled with the glory of God: such is the vision that must guide us at this critical time of Earth's distress, to practical and critical effect. Ignoring this view keeps people of faith and their churches locked into irrelevance while a terrible drama of life and death is being played out in the real world. By contrast, living the ecological vocation in the power of the Spirit sets us off on a great adventure of mind and heart, expanding the repertoire of our love.

The beasts ask of us no less.

NOTES

PREFACE

1   Holmes Rolston III, *Three Big Bangs* (New York: Columbia University Press, 2010), xi.
2   All volumes are published in Cambridge, MA by Harvard University Press:
    Buddhism and Ecology, Mary Evelyn Tucker and Duncan Ryuken Williams, (eds), 1997.
    Confucianism and Ecology, Mary Evelyn Tucker and John Berthrong, (eds), 1998.
    Christianity and Ecology, Dieter Hessel and Rosemary Radford Ruether, (eds), 2000.
    Daoism and Ecology, N. J. Girardot, Liu Xiaogan, and James Miller, (eds), 2001.
    Hinduism and Ecology, Christopher Key Chapple and Mary Evelyn Tucker, (eds), 2000.
    Islam and Ecology, Richard Foltz, Frederick Denny, and Azizan Baharuddin, (eds), 2003.
    Indigenous Traditions, John Grim, ed., 2001.
    Jainism and Ecology, Christopher Key Chapple, ed., 2002.
    Judaism and Ecology, Hava Tirosh-Samuelson, ed., 2002.
    Shinto and Ecology, Rosemarie Bernard, ed., 2003.
3   Aldo Leopold, *A Sand County Almanac and Sketches Here and There* (New York: Oxford University Press, 1949), 110.
4   The website greenfaith.org carries statements from leaders of Catholic, Orthodox and Protestant Christian churches, plus Hindu, Islamic, Buddhist, and Jewish faiths. The National Religious Partnership for the Environment (www.nrpe.org) is a rich source for materials from the U.S. Conference of Catholic Bishops, the National Council of Churches of Christ USA, Evangelical Environmental Network, and Coalition on the Environment and Jewish Life, all of which also have their own websites.

5 Karl Rahner, "Art Against the Horizon of Theology and Piety," *Theological Investigations* Vol. 23 (New York: Crossroad, 1992), 165.

6 Sallie McFague, "An Earthly Theological Agenda," in Carol Adams, ed., *Ecofeminism and the Sacred* (New York: Continuum, 1993), 84–98.

7 Charles Darwin, *On the Origin of Species*, 489. This book went through extensive revisions over the course of its six separate editions. Many scholars consider the first edition to be the cleanest, lacking responses to later criticisms of the book which were current in the 1860s but which by now may be obsolete. In keeping with this consensus, I cite from the first edition, enriched with excellent annotations:

> Charles Darwin, *The Annotated Origin: A Facsimile of the First Edition of On the Origin of Species*, annotated by James Costa (Cambridge, MA and London: Belknap Press of Harvard University Press, 2009).

Wherever a page number is cited after a quotation without further reference, it is from this version of the first edition.

1. BEASTS AND ENTANGLED BANK: A DIALOGUE

1 Paul Santmire, *The Travail of Nature: The Ambiguous Ecological Promise of Christian Theology* (Minneapolis: Fortress, 1985), ch. 3 and 4; Jame Schaefer, *Theological Foundations for Environmental Ethics: Reconstructing Patristic and Medieval Concepts* (Washington, DC: Georgetown University Press, 2009) is a helpful resource compiled by theme.

2 Gordon Kaufman, "A Problem for Theology: The Concept of Nature," *Harvard Theological Review* 65 (1972), 337–66.

3 Christiana Peppard, "Denaturing Nature," *Union Seminary Quarterly Review* 63 (2011), 97–120.

4 Edward Schillebeeckx, "I Believe in God, Creator of Heaven and Earth," in his *God Among Us: The Gospel Proclaimed* (New York: Crossroad, 1983), 93.

5 World Council of Churches, Canberra Assembly, "Giver of Life Sustain Your Creation!," in Michael Kinnamon, ed., *Signs of the Spirit* (Eerdmans: Grand Rapids, MI: 1991), 55.

6 Astute analysis offered by John B. Cobb, *The Earthist Challenge to Economism* (Basingstoke: Macmillan, 1999); and David Korten, "Sustainability and the Global Economy," in Harold Coward and Daniel Maguire, (eds), *Visions of a New Earth* (Albany, NY: State University of New York Press, 2000), 29–42.

7 Ian Barbour, *Religion in an Age of Science* (San Francisco: Harper & Row, 1990) 77–105; revised and expanded as *Religion and Science*, 1997. While this typology refers to the relation between science and religion, many have adapted it to theology understood as an activity done by members of the Christian religion. Others have expanded and nuanced this typology in helpful ways: John Haught, *Science and Religion* (New

York: Paulist, 1995); Niels H. Gregersen and Wentzel van Huyssteen, (eds), *Rethinking Theology and Science: Six Models for the Current Dialogue* (Grand Rapids: Eerdmans, 1998); and Del Ratzsch, "Science and Religion," in Thomas Flint and Michael Rea, (eds), *The Oxford Handbook of Philosophical Theology* (New York: Oxford University Press, 2009), 54–77.

8   William Dembski, *Intelligent Design: The Bridge between Science and Religion* (Downers Grove, Ill.: InterVarsity Press, 1999); and Arthur McCalla, *The Creationist Debate* (New York: Continuum, 2006).

9   Vatican II, "Dogmatic Constitution on Divine Revelation" (*Dei Verbum*), §11.

10   A case in point is Richard Dawkins, *The God Delusion* (New York: Houghton Mifflin, 2006).

11   Kenneth Miller, *Finding Darwin's God* (New York: Harper Collins, 1999), gives a lucid analysis of the blinkered view of both positions from his viewpoint as both a scientist and practicing Catholic.

12   Stephen Jay Gould, *Rocks of Ages* (New York: Ballantine Books, 1999).

13   John Paul II, "Message to the Reverend George Coyne, S.J., Director of the Vatican Observatory," in Robert Russell, William Stoeger, and George Coyne, (eds), *Physics, Philosophy, and Theology: A Common Quest for Understanding* (Vatican City: Vatican Observatory, 1988), M10, M12, and M13.

14   Christopher Mooney, "Theology and Science: A New Commitment to Dialogue," *Theological Studies* 52 (1991), 316.

15   John Paul II, "Message to Coyne," M13.

16   Robert Hughes, *Beloved Dust: Tides of the Spirit in the Christian Life* (New York: Continuum, 2008), 271.

17   Carl Sagan et al., "An Open Letter to the Religious Community," at www.nrpe.org.

18   Edward O. Wilson, *The Creation* (New York: W. W. Norton, 2006), 4.

19   John Paul II, "Truth Cannot Contradict Truth," Address to the Pontifical Academy of Sciences, in Robert Russell, William Stoeger, and Francisco Ayala, (eds), *Evolutionary and Molecular Biology* (Vatican City: Vatican Observatory and Berkeley, CA: Center for Theology and the Natural Sciences, 1998), 4.

20   Carl Safina, *New York Times*, April 30, 2010.

21   John Haught, *God After Darwin: A Theology of Evolution* (Boulder, CO: Westview Press, 2000), chapters 2 and 4.

2. "WHEN WE LOOK ..."

1   In *Origins* the term "races" refers mainly to varieties or kinds of animals and plants, as in races of cabbages, races of pigeons, races of domestic animals. For much of the background material in this chapter I am indebted to *The Cambridge Companion to Darwin*, Jonathan Hodge and Gregory Radick, (eds) (Cambridge: Cambridge University Press, 2009), hereafter *CCD*; and *The Cambridge Companion to the Origin of*

Species, Michael Ruse and Robert Richards, (eds) (Cambridge: Cambridge University Press, 2009), hereafter *CCOS*.

2   Janet Browne's beautifully-written two volume biography tells Darwin's story with acute attention to its social context: *Charles Darwin: Voyaging* (New York: Knopf, 1995); and *Charles Darwin: The Power of Place* (New York: Knopf, 2002); as does Adrian Desmond and James Moore's earlier biography, *Darwin* (New York: Penguin, 1992).

3   Gregory Radick makes the telling point that besides mapping coastlands to extend the empire's reach, the ship was returning three indigenous persons (a fourth had died) to the tip of South America, in the hopes that having been converted to Christianity they would spread this civilizing influence throughout the rest of their group. See his "Is the theory of natural selection independent of its history?" *CCD*, 147–72.

4   First published in London by Henry Colburn in 1839, *The Voyage of the Beagle* has appeared in numerous editions ever since, edited, abridged, annotated, introduced, illustrated, and given appendices by a sequence of fascinated scholars. This and all of Darwin's major publications are available online in a database called the Complete Works of Charles Darwin Online (http://darwin-online.org.uk/).

5   Phillip Sloan, "The making of a philosophical naturalist," *CCD*, 21–43; and Jonathan Hodge, "The notebook programmes and projects of Darwin's London years," *CCD*, 44–72.

6   Darwin's thousands of letters exchanged with over 2,000 correspondents are appearing in another database, The Darwin Correspondence Project (http://www.darwinproject.ac.uk/).

7   Adam Gopnik, *Angels and Ages: A Short Book abut Darwin, Lincoln, and Modern Life* (New York: Knopf, 2009), 95.

8   Janet Browne, *Darwin's Origin of Species: a Biography* (New York: Atlantic Monthly: 2007) is an excellent popular retelling of the story of this book's genesis, writing, and reception; see also Michèle Kohler and Chris Kohler, "The *Origin of Species* as a Book," *CCOS*, 333–51.

9   George Levine, "Foreword," in Gillian Beer, *Darwin's Plots* (Cambridge: Cambridge University Press, 2000), x.

10   A version was finally published in 1975, ed. R. C. Stauffer (Cambridge: Cambridge University Press).

11   David N. Reznick, *The 'Origin' Then and Now: An Interpretive Guide to the 'Origin of Species'* (Princeton: Princeton University Press, 2010), 24.

12   The Alfred Russel Wallace Page contains a full bibliography and the text of many of Wallace's writings (people.wku.edu/charles.smith/index1.htm).

13   Edward O. Wilson, *The Diversity of Life* (Cambridge, MA: Harvard University Press, 2010), 80.

14   Details in Arthur Peacock, "Biological Evolution and Christian Theology – Yesterday and Today," in *Darwinism and Divinity*, ed. John Durant (Oxford: Basil Blackwell, 1985) 101–30.

15 Letter of June 5, 1870, *The letters and diaries of J. H. Newman, XXV*, C. S. Dessain and T. Gornall, (eds) (Oxford: Clarendon Press, 1973), 137.

16 Reznick, *The 'Origin' Then and Now*, 138.

17 Michael Ruse, "The Origin of the *Origin*," *CCOS*, 7–8.

18 Anonymous (Richard Owen), "Darwin on the Origin of Species," *Edinburgh Review*, 3 (1860), 487–532.

19 Hodge, "The notebook programmes," *CCD*, 47.

20 Cited in David Hull, "Darwin's science and Victorian philosophy of science," *CCD*, 182.

21 Ibid. 174.

22 Paley cited in George Levine, "Introduction," in Charles Darwin, *On the Origin of Species* (New York: Barnes and Noble Classics, 2004), xxiii.

23 John Hedley Brooke, "The *Origin* and the Question of Religion," *CCOS*, 259.

24 Charles Darwin, *Autobiography*, edited and with appendix and notes by his grand-daughter Nora Barlow (London: Collins, 1958), 87.

25 The words of nineteenth-century astronomer John Herschel; cited by Costa, *The Annotated 'Origin'*, 352.

26 David Depew, "The Rhetoric of the *Origin of Species*," *CCOS*, 237–55 gives sharp analysis.

27 Darwin, *Voyage of the Beagle*, ch. 1.

28 Cited in Brooke, "The *Origin* and the Question of Religion," *CCOS*, 274.

29 Cited in Brooke, "Darwin and Victorian Christianity," *CCD*, 202.

30 Cited in Sloan, "The making of a philosophical naturalist," *CCD*, 37.

31 Cited in Brooke, "The *Origin* and the Question of Religion," *CCOS*, 271.

32 Hodge, "The notebook programmes," *CCD*, 62.

33 Cited in Brooke, "Darwin and Victorian Christianity," *CCD*, 205.

34 Cited in Costa, *The Annotated Origin*, 472.

35 Cited in Brooke, "Darwin and Victorian Christianity," *CCD*, 215.

36 G.M. Hopkins, "Hurrahing in Harvest," *Gerard Manley Hopkins: The Major Works* (Oxford: Oxford University Press, 1986), 134.

37 Michael Himes, "Finding God in All Things: A Sacramental Worldview and Its Effects," in *As Leaven in the World*, Thomas Landy, ed. (Franklin, WI: Sheed & Ward, 2001), 91–103.

38 Cited in Costa, *The Annotated Origin*, 398.

39 Sallie McFague, *Super, Natural Christians* (Minneapolis: Fortress, 1997), chapters 4 and 5.

3. ENDLESS FORMS MOST BEAUTIFUL

1 Cited in Costa, *The Annotated Origin*, 434.

2 In 1862 Darwin received an orchid from the African island of Madagascar with an astonishingly foot-long throat or nectary. The sweet liquid was held only in the bottom

inch and a half. He ventured to say that Madagascar must be home to an insect with an incredibly long feeding tube, or proboscis, that could reach the nectar while pollinating the plant. Entomologists were dubious, but the predicted giant hawk moth was discovered there in 1903.

3   The diagram reproduced here appears on unnumbered pages between pp. 117 and 118 of the first edition of *Origin*.

4   Wilson, *The Diversity of Life*, 38.

5   Jean Gayon, "From Darwin to today in evolutionary biology," *CCD*, 282.

6   See the Tree of Life project website (tolweb.org), a helpful tool that traces the life history of over 10,000 individual species, from root to branch to the present day.

7   Reznick, *The 'Origin' Then and Now*, 237. Richard Dawkins, *Climbing Mount Improbable* (New York: Norton, 1995), devotes the excellent Chapter 5 to the evolution of the eye, with illustrations, pp. 138–97.

8   Jonathan Weiner, *The Beak of the Finch: A Story of Evolution in Our Time* (Vintage Books: 1995), tells the story with panache.

9   Resnick, *The 'Origin' Then and Now*, 34.

10 Ibid., 381.

11 The Silurian system is a geologic period that extends from around 444 to 416 million years ago. It saw the appearance of bony fish in the sea and small, moss-like plants on land.

12 Browne, *Charles Darwin: The Power of Place*, 62.

4. EVOLUTION OF THE THEORY

1   Arthur Peacocke. *Theology for a Scientific Age* (Minneapolis: Fortress, 1993), 62.

2   Francisco Ayala, "The Evolution of Life: An Overview," in Russell et al., *Evolutionary and Molecular Biology*, 21–57.

3   Diane Paul, "Darwin, social Darwinism and eugenics," *CCD*, 219–45; see also Naomi Beck, "The *Origin* and Political Thought: From Liberalism to Marxism," *CCOS*, 295–313.

4   Darwin, *The Voyage of the Beagle*, ch. 2.

5   Peter Quinn, "The Gentle Darwinians," *Commonweal* (March 9, 2007), 8–16.

6   Online at www.amnh.org/exhibitions/darwin; click on "Evolution Today: Social Darwinism."

7   Wilson, *The Diversity of Life*, 75.

8   Francisco Ayala, "The Theory of Evolution: Recent Successes and Challenges," in Ernan McMullin, ed., *Evolution and Creation* (Notre Dame, IN: University of Notre Dame Press, 1985), 59–90.

9   Reznick, *The 'Origin' Then and Now*, 131.

10 Wilson, *The Diversity of Life*, 148.

11 Details in Robert Richards, "Darwin on mind, morals, and emotions," *CCD*, 96–119.

12 Stephen Jay Gould, *Wonderful Life: The Burgess Shale and the Nature of History* (New York: W. W. Norton, 1989), presents one splendid account.

13 Rolston, *Three Big Bangs*, 3; I have drawn key ideas and examples here from Rolston's book.

14 Ibid.

15 Carl Sagan, *The Dragons of Eden* (New York: Random House, 1977), 11–18.

16 Arthur Peacocke, "Theology and Science Today," in *Cosmos as Creation*, Ted Peters, ed. (Nashville: Abingdon Press, 1989), 32.

17 John Polkinghorne, *The Quantum World* (Princeton: Princeton University, 1984), offers background for the interested non-specialist.

18 James Gleick, *Chaos: Making a New Science* (New York: Penguin, 1987), provides an overall view of the theory; Ilya Prigogine and Isabelle Stengers, *Order Out of Chaos* (New York: Bantam, 1984), offer a key refutation of the idea that chaos amounts to blind, purposeless chance.

19 Term used by John Polkinghorne, "The Laws of Nature and the Laws of Physics," in Robert J. Russell, Nancey Murphy, and C. J. Isham, (eds), *Quantum Cosmology and the Laws of Nature: Scientific Perspectives on Divine Action* (Vatican City: Vatican Observatory and Berkeley: Center for Theology and the Natural Sciences, 1993), 437–48.

20 Paul Davies, *The Cosmic Blueprint* (New York: Simon & Schuster, 1988), 202.

21 Wilson, *The Diversity of Life*, 14.

22 Ibid., 158.

23 Wendell Berry, *The Unsettling of America* (San Francisco: Sierra Club Books, 1977), 86.

24 Brian Swimme and Thomas Berry, *The Universe Story* (San Francisco: HarperSanFrancisco, 1992), 77–8.

25 Wilson, *The Diversity of Life*, 15.

26 Rolston, *Three Big Bangs*, 52, adapted.

27 John Haught, *Making Sense of Evolution: Darwin, God, and the Drama of Life* (Louisville, KY: Westminster John Knox, 2010), 53.

5. THE DWELLING PLACE OF GOD

1 Herbert McCabe, *God, Christ and Us*, Brian Davies, ed. (New York: Continuum, 2003), 103.

2 One of the earliest and most lucid analyses remains Rosemary Radford Ruether, *Sexism and God-Talk* (Boston: Beacon, 1983), ch. 3, "Woman, Body, and Nature: Sexism and the Theology of Creation," 72–92; see her *Gaia and God: An Ecofeminist Theology of Earth Healing* (San Francisco: HarperCollins, 1992). Heather Eaton, *Introducing Ecofeminist Theologies* (London: T&T Clark, 2005), presents the diverse ways other feminist thinkers have mounted this criticism.

3 Carolyn Merchant, *The Death of Nature: Women, Ecology, and the Scientific Revolution* (New York: Harper & Row, 1982), esp. ch. 8, for a superb analysis of Descartes in the context of other modern anti-ecological factors.

4   David Burrell, *Freedom and Creation in Three Traditions* (Notre Dame, IN: University of Notre Dame, 1993), 3–4.

5   Ibid.

6   Anselm Min, *Paths to the Triune God* (Notre Dame, IN: University of Notre Dame, 2005), chs 1 and 2, traces this development and its potential correction.

7   Joseph Sittler, *Evocations of Grace: Writings on Ecology, Theology, and Ethics*, Steven Bouma-Prediger and Peter Bakken, (eds) (Grand Rapids, MI: Eerdmans, 2000), 42.

8   Walter Kasper, *The God of Jesus Christ* (New York: Crossroad, 1984), 202.

9   Ibid., 228.

10   Irenaeus, *Adversus Haereses*, IV.xx.1.

11   David Jensen, "Discerning the Spirit: A Historical Introduction," in David Jensen, ed., *The Lord and Giver of Life* (Louisville: Westminster John Knox, 2008), 9.

12   These images are suggested by Tertullian, *Adversus Praxeas*, 8.

13   Julian of Norwich, *Showings*, Long Text, ch. 59.

14   Augustine, *The Trinity*, VI.1.7 and XV.2.10.

15   Thomas Aquinas, *Summa Contra Gentiles*, IV.19.11.

16   Elizabeth Johnson, *Women, Earth, and Creator Spirit* (New York: Paulist, 1993), 42.

17   J. B. Phillips, *Your God Is Too Small* (New York: Macmillan, 1968), still one of the best popular treatments of how stereotypical images of God fail to do justice to the incomprehensible mystery of the living God.

18   For the number of disciples gathered in the room, see Acts 1.13-15; they numbered 120, including the women disciples and Jesus' family members, not twelve men and the mother of Jesus, as usually depicted in Christian art.

19   Augustine, *Confessions*, VII:7.

20   Jürgen Moltmann, *Spirit of Life* (Minneapolis: Fortress, 1992), 177.

21   *Hildegard of Bingen: Mystical Writings*, Fiona Bowie & Oliver Davies, (eds), (New York: Crossroad, 1990), 91–3.

22   Stephen Hawking, *A Brief History of Time* (New York: Bantam Books, 1988), 174.

23   Cited in Robert Murray, "Holy Spirit as Mother," in *Symbols of Church and Kingdom* (London: Cambridge University Press, 1975), 315.

24   Cited in E. Pataq-Siman, *L'Expérience de l'Esprit d'après la tradition syrienne d'Antioche; Théologie historique* 15 (Paris: Beauchesne, 1971), 155.

25   Augustine, *The Literal Meaning of Genesis*, I.36.

26   Ibid.

27   James D. G. Dunn, *Christology in the Making* (Philadelphia: Westminster, 1980), 163–212.

28   Elisabeth Schüssler Fiorenza, *In Memory of Her* (New York: Crossroad, 1983), 130–40; and Elizabeth Johnson, *She Who Is: The Mystery of God in Feminist Theological Discourse* (New York: Crossroad, 1992), 86–100, provide fuller discussion.

29   Augustine, *The Trinity*, VI.11.

30   Kasper, *The God of Jesus Christ*, 156.

31 Catherine LaCugna, *God For Us: The Trinity and Christian Life* (San Francisco: HarperCollins, 1991), 250.

32 Burrell, *Freedom and Creation*, 31–7.

33 Thomas Aquinas, *Summa Theologiae*, I.8.1.

34 Ibid., I.8.2.

35 Ibid., I.8.1. ad 2.

36 Arthur Peacocke, "Articulating God's Presence in and to the World Uncovered by the Sciences," in Philip Clayton and Arthur Peacocke, (eds), *In Whom We Live and Move and Have Our Being: Panentheistic Reflections on God's Presence in a Scientific World* (Grand Rapids: Eerdmans, 2004), 145. Owen Thomas mounts a strong critique of the term's sloppy usage in "Problems in Panentheism," in Philip Clayton and Zachary Simpson, (eds), *The Oxford Handbook of Religion and Science* (New York: Oxford University Press, 2006), 652–64.

37 Aquinas, *ST*, I.45.3.

38 Ibid.

39 Ibid., I.44.1.

40 W. Norris Clarke, "The Meaning of Participation in St. Thomas," in *Explorations in Metaphysics: Being-God-Person* (Notre Dame, IN: University of Notre Dame, 1994), 89–101.

41 Fran O'Rourke, "Aquinas and Platonism," in Fergus Kerr, ed., *Contemplating Aquinas: On the Varieties of Interpretation* (London: SCM Press, 2003), 265.

42 Aquinas, *ST*, I.19.2

43 Ibid., I.103.2.

44 Ibid., I. 47.1.

45 Denis Edwards, *Ecology at the Heart of Faith* (Maryknoll, NY: Orbis, 2006), 78.

46 Dorothy McDougall, *The Cosmos as Primary Sacrament* (New York: Peter Lang, 2003), and Michael Himes and Kenneth Himes, "Creation and an Environmental Ethic," in *Fullness of Faith: The Public Significance of Theology* (New York: Paulist, 1993), 104–24 develop the sacrament theme.

47 John Haught, *The Promise of Nature* (New York: Paulist, 1993), 76–8.

48 Augustine, "Sermon 68:6," *Sermons* III/3, trans. Edmund Hill (Brooklyn: New City Press, 1991), 225–6.

49 Gerard Manley Hopkins, "God's Grandeur," in *Gerard Manley Hopkins: The Major Works*, Catherine Phillips, ed. (New York: Oxford University Press, 2002), 128.

6. FREE EMPOWERED CREATION

1   Karl Rahner, "Christology in the Setting of Modern Man's Understanding of Himself and His World," *Theological Investigations* Vol. 11 (New York: Seabury, 1974), 225.

2   Irenaeus, *Adversus Haeresis*, 4.20.7, see also 3.20.2 and 5.3.

3 Bernard of Clairvaux, *Bernard of Clairvaux: Treatises III: On Grace and Free Choice, In Praise of the New Knighthood* (Kalamazoo, MI: Cistercian Pub., 1977), ch. 14, #7, p. 106.

4 Karl Rahner, *Foundations of Christian Faith* (New York: Seabury, 1979), 78.

5 Kathryn Tanner, *Christ the Key* (New York: Cambridge University Press, 2010), 275.

6 Wolfhart Pannenberg, *Systematic Theology*, Vol. 1 (Grand Rapids: Eerdmans, 1991), 422.

7 Ian Barbour develops an excellent schema of positions on divine action in *Religion and Science*, 305–32, with chart on 305. See other layouts of models of divine action by Denis Edwards, "Exploring How God Acts," in Philip Rossi, ed., *God, Grace, & Creation* (Maryknoll, NY: Orbis, 2010), 124–46; and Owen Thomas, *God's Activity in the World* (Atlanta, GA: Scholars Press, 1983).

8 Sallie McFague, *The Body of God: An Ecological Theology* (Minneapolis: Fortress, 1993).

9 Aquinas, *ST* I.2., Prologue.

10 David Burrell, "Aquinas' Appropriation of *Liber de causis* to Articulate the Creator as Cause-of-Being," in Kerr, *Contemplating Aquinas*, 82.

11 Aquinas, *SCG* III.69.15; David Burrell, *Aquinas: God and Action* (Notre Dame: University of Notre Dame, 1979), provides extended analysis.

12 Aquinas, *ST* I.22.3.

13 Robin Collins, "Divine Action and Evolution," in Thomas Flint and Michael Rea, (eds), *Philosophical Theology* (New York: Oxford University Press, 2009), 241–61.

14 Aquinas, *ST* I.103.8.

15 Ibid., I.103.6.

16 Ibid., I.105.5.

17 Barbour, *Religion and Science*, 311–12.

18 Peacocke, *Theology for a Scientific Age*, 146.

19 John Polkinghorne, *Science and Theology* (Minneapolis: Fortress, 1998), 86.

20 Schillebeeckx, *God Among Us*, 91.

21 Herbert McCabe, *On Aquinas*, Brian Davies, ed. (New York: Continuum, 2008), 117.

22 Denis Edwards, *The God of Evolution* (New York: Paulist Press, 1999), 47; see his careful development of this idea in *How God Acts* (Minneapolis: Fortress, 2010).

23 Peacocke, *Theology for a Scientific Age*, 118.

24 William Stoeger, "Contemporary Physics and the Ontological Status of the Laws of Nature," in Russell et al., *Quantum Cosmology and the Laws of Nature*, 209–34.

25 Albert Einstein, *Out of My Later Years* (Westport, CT: Greenwood Press, 1970), 61.

26 Cynthia Crysdale and Neil Ormerod, *Creator God, Evolving World* (Minneapolis: Fortress, 2013), describe the statistical view in detail.

27 Strong arguments along these lines are made by influential figures such as Jacques Monod, *Chance and Necessity* (London: Collins, 1972); Gould, *Wonderful Life*; Daniel Dennett, *Darwin's Dangerous Idea: Evolution and the Meaning of Life* (New York: Simon & Schuster, 1995); and Richard Dawkins, *The Blind Watchmaker* (New York: W. W. Norton, 1996).

28 Peacocke, *Theology for a Scientific Age*, 117.

29 Stoeger, "Contemporary Physics and the Ontological Status of the Laws of Nature," 234.

30 Elizabeth Johnson, "Does God Play Dice? Divine Providence and Chance," *Theological Studies* 57 (1996), 3–18.

31 Gould, *Wonderful Life*, 50–2; and William Stoeger, "The Immanent Directionality of the Evolutionary Process and Its Relationship to Teleology," in Russell et al., *Evolutionary and Molecular Biology*, 163.

32 Rolston, *Three Big Bangs*, 32.

33 George Coyne, "Evolution and the Human Person," in Russell, et al., *Evolutionary and Molecular Biology*, 17.

34 Niels Gregersen, "Emergence: What is at Stake for Religious Reflection?" in Philip Clayton and Paul Davies, (eds), *The Re-Emergence of Emergence: The Emergentist Hypothesis from Science to Religion* (New York: Oxford, 2009), 279–302; and Francisco Ayala, "Darwin's Devolution: Design without Designer," in Russell, et al., *Evolutionary and Molecular Biology*, 101–16.

35 Rahner, "Christology within an Evolutionary View of the World," 164. Appreciative critique is offered by Oliver Putz, "Evolutionary Biology in the Theology of Karl Rahner," *Philosophy and Theology* 17 (2002), 85–105.

36 Rahner, "Christology within an Evolutionary View," 166.

37 Ibid., 165.

38 Karl Rahner and Joseph Ratzinger, *Revelation and Tradition* (New York: Herder & Herder, 1966), 12. Michael Petty, *A Faith that Loves the Earth: The Ecological Theology of Karl Rahner* (New York: University Press of America, 1996), parses this in detail.

39 Rahner, "The Secret of Life," *Theological Investigations* Vol. 6 (New York; Seabury, 1974), 149.

40 Ibid., 142.

41 W. Norris Clarke, "Is a Natural Theology Still Possible Today?," in Russell, et al., *Physics, Philosophy, and Theology*, 121. The model of a fugue is developed by Arthur Peacocke, *Theology for a Scientific Age*, 173–7, that of the game by Paul Davies, *God and the New Physics* (New York: Simon & Schuster, 1984); the theme of improvisation is stressed by Peter Geach, *Providence and Evil* (Cambridge: Cambridge University, 1977). Jazz is my suggestion.

42 John Haught, "Darwins Gift to Theology," in Russell et al., *Evolutionary and Molecular Biology*, 415.

43 Dante, *The Divine Comedy: Paradise*, canto 33, line 145.

7. ALL CREATION GROANING

1 Holmes Rolston, "Does Nature Need to be Redeemed?," *Zygon* 29:2 (June 1994) 213.

2 Ibid., 217.

3 Brendan Byrne, "Creation Groaning: An Earth Bible Reading of Romans 8:18-22," in Norman Habel, ed., *Readings from the Perspective of Earth* (Sheffield: Sheffield Academic

Press and Cleveland, OH: Pilgrim Press, 2000), 193–203, includes discussion of the meanings of groaning and this text's critique of anthropocentrism.

4   Holmes Rolston, *Genes, Genesis, and God: Values and Their Origins in Natural and Human History* (Cambridge: Cambridge University Press, 1999), 303.

5   Denis Edwards, "Every Sparrow that Falls to the Ground: The Cost of Evolution and the Christ-Event," *Ecotheology* 11.1 (2006), 106.

6   Peacocke, *Theology for a Scientific Age*, 68–9.

7   John Thiel, *God, Evil, and Innocent Suffering* (New York: Crossroad, 2002), 58.

8   Holmes Rolston, *Science and Religion* (Philadelphia: Temple University Press, 1987), 134.

9   Ibid., 137–40.

10 Terrence Tilley, *The Evils of Theodicy* (Washington, DC: Georgetown University Press, 1991), and Jon Sobrino, *Where Is God?* (Maryknoll, NY: Orbis, 2004). By contrast, Christopher Southgate, *The Groaning of Creation: God, Evolution, and the Problem of Evil* (Louisville, KY: Westminster John Knox, 2008), constructs a theodicy in a christological framework; as does Thomas Tracy, "Evolution, Divine Action, and the Problem of Evil," in Russell, et al., *Evolutionary and Molecular Biology*, 511–30.

11 Celia Deane-Drummond, *Christ and Evolution* (Minneapolis: Fortress, 2009), 172.

12 Jürgen Moltmann, *The Spirit of Life* (Minneapolis: Fortress, 1992), 173–4; 154–5.

13 Jürgen Moltmann, *Sun of Righteousness, Arise!* (Minneapolis: Fortress, 2010), 81.

14 Haught, *Making Sense of Evolution*, 100–1.

15 Gustavo Gutierrez, *The God of Life* (Maryknoll, NY: Orbis, 1991).

16 John Polkinghorne, *Science and Providence: God's Interaction with the World* (London: SPCK, 1989), 66.

17 Ibid., 67.

18 The classic exposition of the *pathos* of God appears in Abraham Heschel, *The Prophets* (New York: Harper & Row, 1962), chs 12–18, see also Terence Fretheim, *The Suffering of God: An Old Testament Perspective* (Philadelphia: Fortress, 1984).

19 Edward Schillebeeckx, *Interim Report on the Books Jesus and Christ* (New York: Crossroad, 1981), 10.

20 This is Edward Schillebeeckx's paraphrase of the central idea of Jesus' preaching: *Jesus: An Experiment in Christology* (New York: Seabury, 1979), 115.

21 Denis Edwards, *Jesus and the Natural World* (Mulgrave: Garratt Pub.: 2012), 33–4.

22 D. Moody Smith, *John* (Nashville: Abingdon Press, 1999), 49.

23 Elizabeth Wainwright, "Which Intertext? A Response to an Ecojustice Challenge: Is Earth Valued in John 1?," in Norman Habel and Vicky Balabanski, (eds), *The Earth Story in the New Testament* (Cleveland: Pilgrim Press, 2002), 83–8, reads the prologue in light of Wisdom poems Sharon Ringe, *Wisdom's Friends: Community and Christology in the Fourth Gospel* (Louisville, KY: Westminster John Knox, 1999), delineates how the Wisdom themes play out in the rest of John's gospel.

24 Rudolf Bultmann, *The Gospel of John: A Commentary* (Philadelphia: Westminster, 1971), 63.

25 Barnabas Lindars, *The Gospel of John* (Grand Rapids, MI: Eerdmans, 1981), 79.

26 Niels Gregersen, "The Cross of Christ in an Evolutionary World," *Dialog: A Journal of Theology* 40 (2001), 192–207.

27 David Toolan, *At Home in the Cosmos* (Maryknoll, NY: Orbis, 2003), 206.

28 Sean McDonagh, *To Care for the Earth* (Santa Fe, NM: Bear and Co., 1986), 118–19; principles of an ecological christology are worked out by Duncan Reid, "Enfleshing the Human: An Earth-Revealing, Earth-Healing Christology," in Denis Edwards, ed., *Earth Revealing, Earth Healing: Ecology and Christian Theology* (Collegeville, MN: Liturgical Press, 2001), 69–83.

29 Rahner, "Christology within an Evolutionary View," 176–7, emended for inclusivity.

30 Karl Rahner, "On the Theology of the Incarnation," *Theological Investigations*, Vol. IV (New York: Seabury, 1974), 116, 118, and passim.

31 Karl Rahner, "The Unity of Spirit and Matter in the Christian Understanding of Faith," *Theological Investigations*, Vol. VI (New York: Crossroad, 1982), 160; emended for inclusivity.

32 John Paul II, encyclical *Lord and Giver of Life (Dominum et Vivificantem)*, §50.

33 Vatican II, *Gaudium et Spes/Pastoral Constitution on the Church in the Modern World*, §22.

34 Teilhard de Chardin, *Hymn of the Universe* (New York: Harper & Row, 1961), 70.

35 Niels Gregersen, ed., *Incarnation and the Depth of Reality* (Minneapolis: Fortress, 2013).

36 Schillebeeckx emphasizes the joy that accompanied Jesus' ministry: *Jesus*, 201 and ff.

37 Lisa Isherwood, "Jesus Past the Posts: An Enquiry into Post-Metaphysical Christology," *Post-Christian Feminisms: A Critical Approach*, Lisa Isherwood and Kathleen McPhillips, (eds), (Surrey: Ashgate, 2008), 206.

38 McFague, *The Body of God*, 161.

39 Burrell, *Freedom and Creation*, 61.

40 Kasper, *The God of Jesus Christ*, 194.

41 Ibid., 195.

42 Jürgen Moltmann, *The Crucified God* (New York: Harper & Row, 1974), 242–6.

43 Benedict XVI, homily at Aosta, July 24, 2009.

44 Ignacio Ellacuría, "The Crucified People," in Ignacio Ellacuría and Jon Sobrino, (eds), *Mysterium Liberationis: Fundamental Concepts of Liberation Theology* (Maryknoll, NY: Orbis, 1993), 580–603.

45 Gregersen, "The Cross of Christ in an Evolutionary World," 205.

46 Arthur Peacocke, "The Cost of New Life," in John Polkinghorne, ed., *The Work of Love: Creation as Kenosis* (Grand Rapids, MI: Eerdmans, 2001), 42.

47 Arlen Gray, privately published meditation, shared in personal correspondence.

48 Southgate, *The Groaning of Creation*, 52.

49 Ibid.; the same conclusion is reached via a different argument by Mark Wallace, "The Wounded Spirit as the Basis for Hope in an Age of Radical Ecology," in Hessler and Ruether, *Christianity and Ecology*, 51–72.

50 Anthony Kelly, *Eschatology and Hope* (Maryknoll, NY: Orbis, 2006), 85.

51 See 1 Cor 15:35-49.

52 Kelly, *Eschatology and Hope*, 94.

53 Rahner, *Foundations of Christian Faith*, 266.

54 Brian Robinette, "Heraclitan Nature and the Comfort of the Resurrection," *Logos* 14:4 (Fall 2011), 25.

55 Ambrose of Milan, *PL* 16:1354.

56 Karl Rahner, "Dogmatic Questions on Easter," *Theological Investigations* Vol. 4 (New York: Seabury, 1974), 129.

57 John Paul II, *Dominum et Vivificantem*, §41.

8. BEARER OF GREAT PROMISE

1   Julian of Norwich, *Showings*, trans. Edmund Colledge and James Walsh (New York: Paulist, 1978), 183.

2   Karl Rahner, in "The Hermeneutics of Eschatological Assertions," *Theological Investigations* Vol. 4 (New York: Seabury: 1974), 323–46, worked out this line of reasoning; a strong consensus has coalesced around it in Catholic theology.

3   Jürgen Moltmann, *God in Creation* (San Francisco: Harper & Row, 1985), 93; Hans Küng, *Eternal Life* (Garden City, NY: Doubleday, 1985), 107–18 makes the same compelling argument.

4   Aquinas, *ST*, I.70.1.

5   Moltmann, *God in Creation*, 276–96, retrieves the meanings of the seventh day of creation, often neglected by theology,

6   For this reading of "*adam* as the earth creature", see Phyllis Trible, "Eve and Adam: Genesis 2–3 Re-read," in Carol Christ and Judith Plaskow, (eds), *Womanspirit Rising* (San Francisco: HarperSanFrancisco, 1979), 74–81.

7   These alternatives are discussed in Gerhard May, *Creatio ex nihilo: 'The Doctrine of 'Creation out of Nothing' in Early Christian Thought* (Edinburgh: T&T Clark, 1994), and David Kelsey, "The Doctrine of Creation from Nothing," in McMullin, *Evolution and Creation*, 176–96.

8   Brian Robinette, "The Difference Nothing Makes: *Creatio ex nihilo*, Resurrection, and Divine Gratuity," *Theological Studies* 73:3 (2011), 529.

9   Augustine, *The Literal Meaning of Genesis* IV. 12, cited with lucid explanation by Ernan McMullin in his introductory essay in *Evolution and Creation*, 1–56.

10 Timothy Ferris, http://wiki.answers.com, "When scientifically will the world end."

11 Kelly, *Eschatology and Hope*, 53–9, traces this aspect of the intelligence of hope in the New Testament; also John Polkinghorne, *The God of Hope and the End of the World* (New Haven, CT: Yale University Press, 2002).

12 Edward Schillebeeckx, *Christ: The Experience of Jesus as Lord* (New York: Seabury, 1980), 468–511, identifies 16 operative metaphors of grace in the New Testament.

13 Vicky Balabansky, "Hellenistic Cosmology and the Letter to the Colossians: Towards an Ecological Hermeneutic," in David Horrell et al., (eds), *Ecological Hermeneutics* (London: T&T Clark, 2010), 94–107.

14 Kallistos Ware, *The Orthodox Church* (Hammondsworth: Penguin, 1963), 239–40.

15 Rahner, *Foundations of Christian Faith*, 284.

16 Rahner, "Christology within an Evolutionary View," 186.

17 Ibid., 184–7.

18 Karl Rahner, "Easter: A Faith that Loves the Earth," in *The Great Church Year: The Best of Karl Rahner's Homilies, Sermons, and Meditations*, Albert Raffelt and Harvey Egan, (eds), (New York: Crossroad, 2001), 195.

19 Ibid.

20 Karl Rahner, "Easter: The Beginning of Glory," in *The Great Church Year*, 191.

21 Rahner, "Easter: A Faith that Loves the Earth," 197.

22 John Muir, "Thoughts on Finding a Dead Yosemite Bear," in Edwin Way Teale, ed., *The Wilderness World of John Muir* (New York: Houghton Mifflin, 2001), 313, 317.

23 Santmire, *The Travail of Nature*, passim.

24 Aquinas, *ST* III *Supplement*, q. 91, a.1; while this part of the *Summa* was composed after Aquinas' death by students who drew on his teaching to complete the work, it is taken to be a fair representation of his thinking.

25 Ibid.

26 Ibid., q. 91, a. 5.

27 Edwards, "Every Sparrow that Falls," 113.

28 Rolston, *Science and Religion*, 133–46.

29 Southgate, *The Groaning of Creation*, sections 1.7, 3.2, 3.3, and 5.4.

30 Jay McDaniel, *Of God and Pelicans* (Louisville, KY: Westminster John Knox, 1989) 45.

31 Edwards, "Every Sparrow that Falls," 119.

32 Ibid. See the careful popular discussion in Jack Wentz, *Will I See My Dog in Heaven?* (Brewster, MA: Paraclete Press, 2009).

33 John Wesley, "The General Deliverance," in Albert Outler, ed., *The Works of John Wesley*, Vol. 2 (Nashville, TN: Abingdon Press, 1985), 446–7.

34 James Dickey, "The Heaven of Animals," in *James Dickey: Selected Poems* (Middletown, CT: Wesleyan University Press, 1998), 31–2.

35 Edwards, "Every Sparrow that Falls," 115.

9. ENTER THE HUMANS

1   Ian Tattersall, *Masters of the Planet* (New York: Palgrave/Macmillan: 2012), presents a chart of the emergence of various human species, xxi–xxii; this book gives an absorbing, elegant account of scientific discoveries of human origins.

2   Camilo Cela-Conde, "The Hominid Evolutionary Journey: A Summary," in Russell et al., *Evolutionary and Molecular Biology*, 71.

3   The larynx's low position in the human throat allows the throat muscles to produce different sounds from the vibrating air column, hence words can be formed. Tatersall delightfully imagines this ability was first discovered and used by children at play; *Masters of the Planet*, 220.

4   Tatersall, *Masters of the Planet*, xii; Wentzel van Huyssteen, *Alone in the World? Human Uniqueness in Science and Theology* (Grand Rapids, MI: Eerdmans, 2006), theorizes that the cave art is associated with religious awareness and shamanistic ritual behavior.

5   Rolston, *Three Big Bangs*, 95.

6   A daunting array of positions is set out in Robert John Russell, Nancey Murphy, Theo C. Meyering, Michael A. Arbib (eds), *Neuroscience and the Person* (Vatican City: Vatican Observatory and Berkeley: Center for Theology and the Natural Sciences, 1999).

7   Tattersall, *Masters of the Planet*, 63.

8   Rolston, *Three Big Bangs*, 98.

9   Tattersall, *Masters of the Planet*, 67.

10   Catherine Keller, "Talk about the Weather: The Greening of Eschatology," in Adams, *Ecofeminism and the Sacred*, 30–49.

11   Bill Mc Kibben, *Eaarth* (sic) (New York: St. Martin's Griffin, 2011), 5, a follow-up to his groundbreaking *The End of Nature* (New York: Random House, 2006).

12   James Nash, *Loving Nature* (Nashville: Abingdon Press, 1991), 44–5.

13   Vatican II, *Gaudium et Spes*, par. 50.

14   John Paul II, General Audience address, September 5, 1984.

15   Jared Diamond, *Collapse* (New York: Penguin Press, 2005), documents this pattern over a spread of different civilizations.

16   Nash, *Loving Nature*, 41.

17   Ibid.

18   David Loy, "The Religion of the Market," in Coward and Maguire, *Visions of a New Earth*, 15–28, makes this analogy with piercing accuracy.

19   Strong links between social and ecological justice are forged from the perspective of liberation theology by Leonardo Boff & Virgilio Elizondo, (eds), *Ecology and Poverty, Concilium* 1995/5 (Maryknoll, NY: Orbis, 1995); from feminist theology by Rosemary Radford Ruether, ed., *Women Healing Earth* (Maryknoll, NY: Orbis, 1996); and from economic and legal theory by Paolo Galizzi, ed., *The Role of the Environment in Poverty Alleviation* (New York: Fordham University Press, 2008).

20   Recounted in Wangari Maathai, *Unbowed: A Memoir* (New York: Knopf, 2006).

21   Nash, *Loving Nature*, 27.

22   John Waldman, *Heartbeats in the Muck* (New York: Fordham University Press, 2011) 20.

23   Ibid., 21.

24   Ibid., 22.

25 Sallie McFague, *Blessed Are the Consumers* (Minneapolis: Fortress, 2013), 22.

26 Elizabeth Kolbert, "Recall of the Wild," *The New Yorker* (December 24 and 31, 2012), 56.

27 www.unep.org.

28 Wilson, *The Diversity of Life*, 280.

29 See www.iucn.org (International Union for Conservation of Nature) to keep track of lists of extinct and endangered species.

30 Sandra Blakeslee, *New York Times* (December 25, 2012), D3.

31 Carl Zimmer, "Apes," *New York Times Book Review* (June 2, 2013), 34.

32 John Feehan, *The Singing Heart of the World: Creation, Evolution, and Faith* (Maryknoll, NY: Orbis Press, 2012), 77.

33 Jonathan Schell, *The Fate of the Earth* (New York: Avon Books, 1982), 117.

34 Wilson, *The Diversity of Life*, 15.

35 Ibid., 74.

36 Lydia Millet, "The Child's Menagerie," *New York Times*, December 8, 2012.

37 John Haught, *The Promise of Nature;* 122, and his *The Cosmic Adventure* (New York: Paulist Press, 1984).

38 Haught, *Making Sense of Evolution*, 76.

39 Ibid.

40 Ibid. 52.

41 William French, "Subject-centered and Creation-centered Paradigms in Recent Catholic Thought," *The Journal of Religion* 70 (1990), 71–2.

42 Catholic Bishops of the Philippines, "What Is Happening to Our Beautiful Land? A Pastoral Letter on Ecology," in Drew Christiansen and Walter Grazer, (eds), *"And God Saw That It Was Good": Catholic Theology and the Environment* (Washington, DC: U.S. Catholic Conference, 1996), 316.

43 John Paul II, "Peace with God the Creator, Peace with All of Creation" (January 1, 1990), #15; also published as "The Ecological Crisis: A Common Responsibility;" italics in the original.

44 Ibid. #7.

45 Ibid. #16.

46 Bartholomew, "Encyclical of His All-Holiness for the Church New Year," September 1, 2012.

47 Albert Einstein, cited in Michael Dowd, *Earthspirit* (Mystic, CT: Twenty-Third Pub., 1991), 81.

48 Larry Rasmussen, *Earth Community, Earth Ethics* (Maryknoll, NY: Orbis Press, 1996), 4.

49 Edwards, *How God Acts*, 135.

10. THE COMMUNITY OF CREATION

1   Most famously, Lynn White, "The Historical Roots of Our Ecologic Crisis," *Science* 155 (March 10, 1967), 1203–7.

2   Richard Bauckham, *The Bible and Ecology: Rediscovering the Community of Creation* (Waco, TX: Baylor University Press, 2010), 19.

3   Richard Clifford, "Genesis 1–3: Permission to Exploit Nature?" *The Bible Today* 26:3 (May 1988), 135.

4   Keith Carley, "Psalm 8: An Apology for Domination," in Habel, *Readings from the Perspective of Earth*, 111–24; and Norman Habel, "'Is the Wild Ox Willing to Serve You?': Challenging the Mandate to Dominate," in Norman Habel & Shirley Wurst, (eds), *The Earth Story in Wisdom Traditions* (Cleveland, OH: Pilgrim Press, 2001), 179–89.

5   Bauckham, *The Bible and Ecology*, 37.

6   See the superb discussion by Calvin DeWitt, *Earth-Wise: A Guide to Hopeful Creation Care*, 3rd edn (Grand Rapids: FaithAlive, 2011); and the journal *Creation Care*.

7   Critiques are discussed at length by Ernst Conradie, "What on Earth is an Ecological Hermeneutics?," in Horrell, *Ecological Hermeneutics*, 295–313.

8   Bauckham, *Bible and Ecology*, 64.

9   Celia Deane-Drummond and David Clough, *Creaturely Theology; On God, Humans, and Other Animals* (London: SCM Press, 2009), 1–2.

10 Ibid., 45.

11 This is the compelling interpretation by Gustavo Gutiérrez, *On Job: God-Talk and the Suffering of the Innocent* (Maryknoll: Orbis, 1987), 82–92.

12 Clare Palmer, "Stewardship: A Case Study in Environmental Ethics," in Ian Ball et al., (eds), *The Earth Beneath: A Critical Guide to Green Theology* (London: SPCK, 1992), 70.

13 Bauckham, *The Bible and Ecology*, 132.

14 McFague, *Blessed Are the Consumers*, 23.

15 Bill Mc Kibben, *The Comforting Whirlwind: God, Job, and the Scale of Creation* (Grand Rapids: Eerdmans, 1994), 63.

16 Bauckham, *Bible and Ecology*, 69. Arthur Walker-Jones, *The Green Psalter: Resources for an Ecological Spirituality* (Minneapolis: Fortress, 2009), 111–64, demonstrates how the creation psalms are intertwined with a vision of justice in society: ecojustice establishes harmony in all realms.

17 These lovely metaphors appear in patristic and medieval sources: Basil of Caesarea likened the praise of creatures to a chorus; John of the Cross contemplated it as a symphony; see texts in Schaefer, *Theological Foundations for Ecological Ethics*, 103–20.

18 A round of criticism vs. defense of the idea of Earth's voice appears in "The Voice of Earth: More then Metaphor?" in Norman Habel, ed. *The Earth Story in the Psalms and the Prophets* (Sheffield: Sheffield Academic Press and Cleveland, Ohio: Pilgrim Press, 2001), 23–8.

19 Augustine, *On the Psalms*, Ancient Christian Writers Vol. 29 (New York: Newman Press, 1960), 272.

20 Bauckham, *Bible and Ecology*, 79.

21 Daniel Hardy and David Ford, *Jubilate: Theology in Praise* (London: Darton, Longman & Todd, 1984), 82.

22 William Urbrock, "The Earth Song in Psalms 90–92," in Habel, *The Earth Story in the Psalms and the Prophets*, 65–83, works out a strong terracentric hermeneutic of voice.

23 Abraham Heschel, *Man's Quest for God* (New York: Scribner, 1954), 82.

24 Bauckham, *Bible and Ecology*, 102.

25 John Paul II, "Peace with God the Creator, Peace with All of Creation," #16.

26 Brian Patrick, cited in Dowd, *Earthspirit*, 40.

27 Erin Lothes Biviano, "Worldviews on Fire: Understanding the Inspiration for Congregational Religious Environmentalism," *Cross Currents* 64 (December 2012), 495–511, analyzes the motivation for ecological action across religions; David Rhoads, ed., *Earth and Word: Classic Sermons on Saving the Planet* (New York: Continuum, 2007), contains fabulous examples attuned to local places; Dieter Hessel, ed., *Theology for Earth Community: A Field Guide* (Maryknoll, NY: Orbis Press, 1996), presents a spread of examples of church action and inaction for ecojustice.

28 John Calvin, *Opera Selecta* 9.793.795; cited by Francois Wendel, *Calvin: The Origins and Development of His Thought* (New York: Harper & Row, 1963), 34.

29 McFague, *The Body of God*, 200–2.

30 Roseanne Murphy, *Martyr of the Amazon: The Life of Sister Dorothy Stang* (Maryknoll, NY: Orbis, 2007).

SELECT BIBLIOGRAPHY

Adams, Carol, ed. *Ecofeminism and the Sacred*. New York: Continuum, 1993.

Aquinas, Thomas. *Summa Theologiae*. New York: Benziger, 1947.

Augustine. *The Literal Meaning of Genesis*, Vol. I. New York: Paulist, 1982.

Ayala, Francisco. "The Theory of Evolution: Recent Successes and Challenges," in McMullin, 1985: 59–90.

—"The Evolution of Life: An Overview," in Russell, Stoeger, and Ayala, 1998: 21–57.

Barbour, Ian. *Religion in an Age of Science*. San Francisco: Harper & Row, 1990; revised and expanded as *Religion and Science*, 1997.

Bartholomew, Ecumenical Patriarch. "Encyclical of His All-Holiness for the Church New Year," Sept. 1, 2012.

Bauckham, Richard. *The Bible and Ecology: Rediscovering the Community of Creation*. Waco, TX: Baylor University Press, 2010.

Beck, Naomi. "The *Origin* and Political Thought: From Liberalism to Marxism," in Ruse and Richards, 2009: 295–313.

Beer, Gillian. *Darwin's Plots*. Cambridge: Cambridge University Press, 2000.

Berry, Thomas. *The Dream of the Earth*. San Francisco: Sierra Club Books, 1988.

Biviano, Erin Lothes. "Worldviews on Fire: Understanding the Inspiration for Congregational Religious Environmentalism," *Cross Currents* 62/4 (December 2012): 495–511.

Boff, Leonardo and Virgilio Elizondo (eds), *Ecology and Poverty*. Maryknoll: Orbis, 1995.

Brooke, John Hedley. "Darwin and Victorian Christianity," in Hodge and Radick, 2009: 197–218.

—"'Laws impressed on matter by the Creator'? The *Origin* and the Question of Religion," in Ruse and Richards, 2009: 256–74.

Browne, Janet. *Charles Darwin: A Biography*, Vol. 1, *Voyaging*. New York: Knopf, 1995.

—*Charles Darwin: A Biography*, Vol. 2, *The Power of Place*. New York: Knopf, 2002.

—*Darwin's Origin of Species: A Biography*. New York: Atlantic Monthly Press, 2007.

Burrell, David. *Aquinas: God and Action*. Notre Dame, IN: University of Notre Dame Press, 1979.

—*Freedom and Creation in Three Traditions*. Notre Dame, IN: University of Notre Dame Press, 1993.

Byrne, Brendan. "Creation Groaning: An Earth Bible Reading of Romans 8:18-22," in Habel, 2000: 193–203.

Camosy, Charles. "Non-Human Animals," in his *Peter Singer and Christian Ethics: Beyond Polarization*. Cambridge: Cambridge University Press, 2012: 88–136.

Carley, Keith. "Psalm 8: An Apology for Domination," in Habel, 2000: 111–24.

Cela-Conde, Camilo. "The Hominid Evolutionary Journey: A Summary," in Russell, Stoeger, and Ayala, 1998: 59–78.

Christiansen, Drew and Walter Grazer (eds), *"And God Saw That It Was Good": Catholic Theology and the Environment*. Washington, DC: U.S. Catholic Conference, 1996.

Clarke, W. Norris. "Is a Natural Theology Still Possible Today?", in Russell, Stoeger, and Coyne, 1988: 103–23.

—*Explorations in Metaphysics: Being-God-Person*. Notre Dame, IN: University of Notre Dame Press, 1994.

Clayton, Philip and Arthur Peacocke (eds), *In Whom We Live and Move and Have Our Being: Panentheistic Reflections on God's Presence in a Scientific World*. Grand Rapids: Eerdmans, 2004.

—and Zachary Simpson (eds), *The Oxford Handbook of Religion and Science*. New York: Oxford University Press, 2006.

—and Paul Davies (eds), *The Re-Emergence of Emergence: The Emergetist Hypothesis from Science to Religion*. New York: Oxford University Press, 2009.

Cobb, John. *The Earthist Challenge to Economism*. Basingstoke: MacMillan, 1999.

Coward, Harold and Daniel Maguire (eds), *Visions of a New Earth: Religious Perspectives on Population, Consumption, and Ecology*. Albany: State University of New York, 2000.

Coyne, George. "Evolution and the Human Person," in Russell, Stoeger, and Ayala, 1998: 11–17.

Crysdale, Cynthia and Neil Ormerod. *Creator God, Evolving World*. Minneapolis: Fortress, 2013.

Darwin, Charles. *Voyage of the Beagle*. Janet Browne and Michael Neve (eds), London: Penguin, 1989.

—*The Annotated Origin: A Facsimile of the First Edition of On the Origin of Species*. Annotated by James Costa. Cambridge, MA: Belknap Press of Harvard University Press, 2009.

—*Autobiography*. Nora Barlow, ed. London: Collins, 1958.

Davies, Paul. *God and the New Physics*. New York: Simon & Schuster, 1984.

Dawkins, Richard. *Climbing Mount Improbable*. New York: W. W. Norton, 1995.

Deane-Drummond, Celia. *Christ and Evolution*. Minneapolis: Fortress, 2009.

—and David Clough (eds), *Creaturely Theology: On God, Humans, and Other Animals*. London: SCM Press, 2009.

Dembski, William and James Kushiner (eds), *Signs of Intelligence: Understanding Intelligent Design*. Grand Rapids: Brazos Press, 2001.

Dennett, Daniel. *Darwin's Dangerous Idea: Evolution and the Meaning of Life*. New York: Simon & Schuster, 1995.

Depew, David. "The Rhetoric of the *Origin of Species*," in Ruse and Richards, 2009: 237–55.

DeWitt, Calvin. *Earth-Wise: A Guide to Hopeful Creation Care*. Grand Rapids: FaithAlive, 2011.

Diamond, Jared. *Collapse*. New York: Penguin, 2005.

Dowd, Michael. *Earthspirit: A Handbook for Nurturing an Ecological Christianity*. Mystic, CT: Twenty-Third Publications, 1991.

Dunn, James D. G. *Christology in the Making*. Philadelphia: Westminster, 1980.

Durant, John, ed. *Darwinism and Divinity*. Oxford: Basil Blackwell, 1985.

Durrwell, F. X. *Holy Spirit of God: An Essay in Biblical Theology*. London: Geoffrey Chapman, 1986.

Eaton, Heather. *Introducing Ecofeminist Theologies*. London: T&T Clark, 2005.

Edwards, Denis. *The God of Evolution*. New York: Paulist, 1999.

—"Every Sparrow that Falls to the Ground: The Cost of Evolution and the Christ-Event," *Ecotheology* 11/1 (2006): 103–23.

—*How God Acts: Creation, Redemption, and Special Divine Action*. Minneapolis: Fortress, 2010.

—ed. *Earth Revealing, Earth Healing: Ecology and Christian Theology*. Collegeville, MN: Liturgical Press, 2001.

Einstein, Albert. *Out of My Later Years*. Westport, CT: Greenwood Press, 1970.

Ellacuría, Ignacio. "The Crucified People," in *Mysterium Liberationis: Fundamental Concepts of Liberation Theology*. Ignacio Ellacuría and Jon Sobrino (eds), Maryknoll: Orbis, 1993: 580–603.

Ferris, Timothy. *The Whole Shebang: A State-of-the-Universe Report*. New York: Simon and Schuster, 1997.

French, William. "Subject-centered and Creation-centered Paradigms in Recent Catholic Thought," *The Journal of Religion* 70/1 (1990): 48–72.

Fretheim, Terence. "Nature's Praise of God in the Psalms," *Ex Auditu* 3 (1987): 16–30.

Galizzi, Paolo. *The Role of the Environment in Poverty Alleviation*. New York: Fordham University Press, 2008.

Gebara, Ivone. *Longing for Running Water: Ecofeminism and Liberation*. Minneapolis: Fortress 1999.

Gleick, James. *Chaos: Making a New Science*. New York: Penguin, 1987.

Goodenough, Ursula. *The Sacred Depths of Nature*. New York: Oxford University Press, 1998.

Gopnik, Adam. *Angels and Ages: A Short Book about Darwin, Lincoln, and Modern Life*. New York: Knopf, 2009.

Gould, Stephen Jay. *Wonderful Life: The Burgess Shale and the Nature of History*. New York: W. W. Norton, 1989.

—*Rocks of Ages*. New York: Ballantine Books, 1999.

Gregersen, Niels. "The Cross of Christ in an Evolutionary World," *Dialog: A Journal of Theology* 40 (2001): 192–207.

—and Wentzel van Huyssteen, (eds), *Rethinking Theology and Science: Six Models for the Current Dialogue*. Grand Rapids: Eerdmans, 1998.

Gustafson, James. *Ethics from a Theocentric Perspective*, Vol. II. Chicago: University of Chicago Press, 1984.

Gutiérrez, Gustavo. *On Job: God-Talk and the Suffering of the Innocent*. Maryknoll: Orbis, 1987.

—*The God of Life*. Maryknoll: Orbis, 1991.

Habel, Norman, ed. *Readings from the Perspective of Earth*. Sheffield: Sheffield Academic Press and Cleveland, OH: Pilgrim Press, 2000.

—ed. *The Earth Story in the Psalms and the Prophets*. Sheffield: Sheffield Academic Press and Cleveland, OH: Pilgrim Press, 2001.

—and Shirley Wurst, (eds), *The Earth Story in Wisdom Traditions*. Sheffield: Sheffield Academic Press and Cleveland, OH: Pilgrim Press, 2001.

Hallman, David, ed. *Ecotheology: Voices from South and North*. Geneva: WCC Publications and Maryknoll: Orbis Press, 1994.

Hansell, Mike. *Built by Animals*. Oxford: Oxford University Press, 2009.

Haught, John. *The Promise of Nature*. New York: Paulist, 1993.

—"Ecology and Eschatology," in Christiansen and Grazer, 1996: 47–64.

—*God After Darwin: A Theology of Evolution*. Boulder, CO: Westview Press, 2000.

—*Making Sense of Evolution: Darwin, God, and the Drama of Life*. Louisville: Westminster John Knox, 2010.

Hawking, Stephen. *A Brief History of Time*. New York: Bantam Books, 1988.

Hessel, Dieter and Rosemary Radford Ruether, (eds), *Christianity and Ecology*. Cambridge, MA: Harvard University Press, 2000.

Hildegard of Bingen. *Mystical Writings*. New York: Crossroad, 1990.

Hilkert, Mary Catherine. "Preaching from the Book of Nature," *Worship* 76/4 (July 2002): 290–313.

Himes, Michael. "Finding God in All Things: A Sacramental Worldview and Its Effects," in Thomas Landy, ed. *As Leaven in the World*. Franklin, WI: Sheed & Ward, 2001: 91–103.

—and Kenneth Himes. "Creation and an Environmental Ethic," in their *Fullness of Faith: The Public Significance of Theology*. New York: Paulist, 1993: 104–24.

Hinze, Christine. "Catholic Social Teaching and Ecological Ethics," in Christiansen and Grazer, 1996: 165–82.

Hodge, Jonathan and Gregory Radick (eds), *The Cambridge Companion to Darwin*. Cambridge: Cambridge University Press, 2009.

Hopkins, G. M. *Gerard Manley Hopkins: The Major Works*. Oxford: Oxford University Press, 1986.

Horrell, David, Cherryl Huny, Christopher Southgate and Francesca Stavrakopoulou, (eds), *Ecological Hermeneutics*. London: T&T Clark, 2010.

Hughes, Robert. *Beloved Dust: Tides of the Spirit in the Christian Life*. New York: Continuum, 2008.

Hull, David. "Darwin's science and Victorian philosophy of science," in Hodge and Radick, 2009: 173–96.

van Huyssteen, Wentzel. *Alone in the World? Human Uniqueness in Science and Theology*. Grand Rapids: Eerdmans, 2006.

Jensen, David, ed. *The Lord and Giver of Life*. Louisville: Westminster John Knox, 2008.

John Paul II, Pope. "Message to the Reverend George Coyne, S.J., Director of the Vatican Observatory," in Russell, Stoeger, and Coyne, 1988: M1–M14.

—"Peace with God the Creator, Peace with All Creation," World Day of Peace Message January 1, 1990, in Christiansen and Grazer, 1996: 215–22.

—"Truth Cannot Contradict Truth," address to the Pontifical Academy of Sciences, in Russell, Stoeger, and Ayala, 1998: 2-9.

Johnson, Elizabeth. *She Who Is: The Mystery of God in Feminist Theological Discourse*. New York: Crossroad, 1992.

—*Women, Earth, and Creator Spirit*. New York: Paulist, 1993.

—"Does God Play Dice? Divine Providence and Chance," *Theological Studies* 57 (1996): 3–18.

Julian of Norwich. *Showings*. New York: Paulist, 1978.

Kasper, Walter. *The God of Jesus Christ*. New York: Crossroad, 1984.

Kaufman, Gordon. "A Problem for Theology: The Concept of Nature," *Harvard Theological Review* 65 (1972): 337–66.

Keller, Catherine. "Talk about the Weather: The Greening of Eschatology," in Adams, 1993: 30–49.

Kelly, Anthony. *Eschatology and Hope*. Maryknoll: Orbis, 2006.

Kohler, Michèle and Chris Kohler, "The *Origin of Species* as a Book," in Ruse and Richards, 2009: 333–51.

Kolbert, Elizabeth. *Field Notes from a Catastrophe: Man, Nature, and Climate Change*. New York: Bloomsbury, 2006.

Korten, David. "Sustainability and the Global Economy," in Coward and Maguire, 2000: 29–42.

Küng, Hans. *Eternal Life*. Garden City, NY: Doubleday, 1985.

LaCugna, Catherine. *God For Us: The Trinity and Christian Life*. San Francisco: HarperCollins, 1991.

Levine, George. "Introduction," in Charles Darwin, *On the Origin of Species*. New York: Barnes and Noble Classics, 2004: xiii–xxxiv.

Loy, David. "The Religion of the Market," in Coward and Maguire, 2000: 15–28.

Maathai, Wangari. *Unbowed: A Memoir.* New York: Knopf, 2006.

Macquarrie, John. *Principles of Christian Theology*. New York: Scribner's, 1977.

May, Gerhard. *Creatio ex nihilo: The Doctrine of 'Creation out of Nothing' in Early Christian Thought*. Edinburgh: T&T Clark, 1994.

McCabe, Herbert. *God, Christ and Us*, ed. Brian Davies. New York: Continuum, 2003.

—*On Aquinas*. New York: Continuum, 2008.

McCalla, Arthur. *The Creationist Debate*. New York: Continuum, 2006.

McDaniel, Jay. *Of God and Pelicans*. Louisville: Westminster John Knox, 1989.

McDonagh, Sean. *To Care for the Earth*. Santa Fe: Bear and Co., 1986.

McFague, Sallie. "An Earthly Theological Agenda," in Adams, 1993: 84–98.

—*The Body of God: An Ecological Theology*. Minneapolis: Fortress, 1993.

—*Blessed Are the Consumers: Climate Change and the Practice of Restraint*. Minneapolis: Fortress, 2013.

McKibben. Bill. *The End of Nature*. New York: Random House, 2006.

—*Eaarth* (sic). New York: St. Martin's Griffin, 2011.

McMullin, Ernan, ed. *Evolution and Creation*. Notre Dame, IN: University of Notre Dame Press 1986.

Merchant, Carolyn. *The Death of Nature: Women, Ecology, and the Scientific Revolution*. New York: Harper & Row, 1982.

Miller, Kenneth. *Finding Darwin's God*. New York: Harper Collins, 1999.

Min, Anselm. *Paths to the Triune God*. Notre Dame, IN: University of Notre Dame Press, 2005.

Moltmann, Jürgen. *The Crucified God*. New York: Harper & Row, 1974.

—*Spirit of Life*. Minneapolis: Fortress, 1992.

—*God in Creation*. Minneapolis: Fortress, 1993.

Mooney, Christopher. "Theology and Science: A New Commitment to Dialogue," *Theological Studies* 52 (1991): 289–322.

Muir, John. *My First Summer in the Sierra*. Boston: Houghton Mifflin, 1998.

Murphy, Roseanne. *Martyr of the Amazon: The Life of Sister Dorothy Stang*. Maryknoll: Orbis, 2007.

Murray, Robert. *Symbols of Church and Kingdom: A Study in Early Syriac Tradition*. London: Cambridge University Press, 1975.

Nash, James. *Loving Nature*. Nashville: Abingdon Press, 1991.

O'Rourke, Fran. "Aquinas and Platonism," in Fergus Kerr, ed., *Contemplating Aquinas: On the Varieties of Interpretation*. London: SCM Press, 2003.

Pannenberg, Wolfhart. *Systematic Theology*, Vol. 1. Grand Rapids: Eerdmans, 1991.

Paul, Diane. "Darwin, Social Darwinism and Eugenics," in Hodge and Radick, 2009: 219–45.

Peacock, Arthur. *God and the New Biology*. San Francisco: Harper & Row, 1986.

—*Theology for a Scientific Age*. Minneapolis: Fortress, 1993.

—"The Cost of New Life," in Polkinghorne, 2001: 21–42.

Peppard, Christiana. "Denaturing Nature," *Union Seminary Quarterly Review* 63 (2011): 97–120.

Peters, Ted, ed. *Cosmos as Creation*. Nashville: Abingdon Press, 1989.

Petty, Michael. *A Faith that Loves the Earth: The Ecological Theology of Karl Rahner*. New York: University Press of America, 1996.

Phillips, J. B. *Your God Is Too Small.* New York: MacMillan, 1968.

Placher, William. *Narratives of a Vulnerable God.* Louisville: Westminster John Knox, 1994.

Polkinghorne, John. *Science and Providence: God's Interaction with the World.* London: SPCK, 1989.

—ed. *The Work of Love: Creation as Kenosis.* Grand Rapids: Eerdmans, 2001.

—*The God of Hope and the End of the World.* New Haven: Yale University Press, 2002.

Quinn, Peter. "The Gentle Darwinians," *Commonweal* (Mar. 9, 2007): 8–16.

Radick, Gregory. "Is the theory of natural selection independent of its history?", in Hodge and Radick, 2009: 147–72.

Rahner, Karl. "The Hermeneutics of Eschatological Statements," *Theological Investigations*, Vol. 4. New York: Seabury: 1974: 323–46.

—*The Trinity.* New York: Seabury, 1974.

—"Christology within an Evolutionary View of the World," *Theological Investigations*, Vol. 5. New York: Seabury, 1975: 157–92.

—*Foundations of Christian Faith.* New York: Crossroad, 1978.

—"The Unity of Spirit and Matter in the Christian Understanding of Faith," *Theological Investigations*, Vol. 6. New York: Crossroad, 1982: 153–77.

—*The Great Church Year: The Best of Karl Rahner's Homilies, Sermons, and Meditations*, Albert Raffelt, ed. New York: Crossroad, 1993.

Rasmussen, Larry. *Earth Community, Earth Ethics.* Maryknoll: Orbis, 1997.

Reznick, David. *The 'Origin' Then and Now: An Interpretive Guide to the 'Origin of Species'.* Princeton: Princeton University Press, 2010.

Rhoads, David, ed. *Earth and Word: Classic Sermons on Saving the Planet.* New York: Continuum, 2007.

Ringe, Sharon. *Wisdom's Friends: Community and Christology in the Fourth Gospel.* Louisville: Westminster John Knox, 1999.

Robinette, Brian. *Grammars of Resurrection: A Christian Theology of Presence and Absence.* New York: Crossroad, 2009.

—"The Difference Nothing Makes: *Creatio ex nihilo*, Resurrection, and Divine Gratuity," *Theological Studies* 72/3 (September 2011): 525–57.

Rolston, Holmes. *Science and Religion*. Philadelphia: Temple University Press, 1987.

—"Does Nature Need to be Redeemed?," *Zygon* 29/2 (June 1994): 205–29.

—*Three Big Bangs*. New York: Columbia University Press, 2010.

Rossi, Philip, ed. *God, Grace, & Creation*. Maryknoll: Orbis, 2010.

Ruether, Rosemary Radford. *Sexism and God-Talk*. Boston: Beacon, 1983.

—*Gaia and God: An Ecofeminist Theology of Earth Healing*. San Francisco: HarperCollins, 1992.

—ed. *Women Healing Earth*. Maryknoll: Orbis, 1996.

Ruse, Michael and Robert Richards (eds), *The Cambridge Companion to the 'Origin of Species'*. Cambridge: Cambridge University Press, 2009.

Russell, Robert J., William Stoeger, and George Coyne (eds), *Physics, Philosophy, and Theology*. Vatican City: Vatican Observatory, 1988.

—Nancey Murphy, and C. J. Isham (eds), *Quantum Cosmology and the Laws of Nature*. Vatican City: Vatican Observatory and Berkeley: Center for Theology and the Natural Sciences, 1993.

—William Stoeger, and Francisco Ayala (eds), *Evolutionary and Molecular Biology*. Vatican City: Vatican Observatory, and Berkeley: Center for Theology and the Natural Sciences, 1998.

—Michael Arbib, Theo C. Meyering, and Nancey Murphy (eds), *Neuroscience and the Person*. Vatican City: Vatican Observatory, and Berkeley: Center for Theology and the Natural Sciences, 2000.

Sagan, Carl. *Cosmos*. New York: Ballantine Books, 1980.

Santmire, Paul. *The Travail of Nature: The Ambiguous Promise of Christian Theology*. Minneapolis: Fortress, 1985.

Schaab, Gloria. *Trinity in Relation: Creation, Incarnation, and Grace in an Evolving Cosmos*. Winona, MI: Anselm Academic, 2012.

Schaefer, James. *Theological Foundations for Ecological Ethics*. Washington, DC: Georgetown University Press, 2009.

Schell, Jonathan. *The Fate of the Earth*. New York: Avon Books, 1982.

Schillebeeckx, Edward. *Jesus: An Experiment in Christology*. New York: Seabury, 1979.

—*God Among Us*. New York: Crossroad, 1983.

Sobrino, Jon. *Where Is God?* Maryknoll: Orbis, 2004.

Southgate, Christopher. *The Groaning of Creation: God, Evolution, and the Problem of Evil*. Louisville: Westminster John Knox, 2008.

Stoeger, William. "Contemporary Physics and the Ontological Status of the Laws of Nature," in Russell, Murphy, and Isham, 1993: 209–34.

Swimme, Brian and Thomas Berry. *The Universe Story: From the Primordial Flaring Forth to the Ecozoic Era.* San Francisco: HarperSanFrancisco, 1992.

Tanner, Kathryn. *Christ the Key.* New York: Cambridge University Press, 2010.

Tatersall, Ian. *Masters of the Planet.* New York: Palgrave Macmillan, 2012.

Teilhard de Chardin, Pierre. *Hymn of the Universe.* New York: Harper & Row, 1961.

Thiel, John. *God, Evil, and Innocent Suffering.* New York: Crossroad, 2002.

Thomas, Owen. *God's Activity in the World.* Atlanta: Scholars Press, 1983.

—"Problems in Panentheism," in Clayton and Simpson, 2006: 652–64.

Tilley, Terrence. *The Evils of Theodicy.* Washington, DC: Georgetown University Press, 1991.

Toolan, David. *At Home in the Cosmos.* Maryknoll: Orbis, 2003.

Trible, Phyllis. *God and the Rhetoric of Sexuality.* Philadelphia: Fortress Press, 1978.

Waldman, John. *Heartbeats in the Muck.* New York: Fordham University Press, 2011.

Walker-Jones, Arthur. *The Green Psalter: Resources for an Ecological Spirituality.* Minneapolis: Fortress, 2009.

Wallace, Mark. "The Wounded Spirit as the Basis for Hope in an Age of Radical Ecology," in Hessel and Ruether, 2000: 51–72.

Weiner, Jonathan. *The Beak of the Finch: A Story of Evolution in Our Time.* New York: Vintage Books, 1995.

Wesley, John. "The General Deliverance," in *The Works of John Wesley,* Vol. 2. Albert Outler, ed. Nashville: Abingdon, 1985: 436–50.

White, Lynn. "The Historical Roots of Our Ecologic Crisis," *Science* 155 (March 10, 1967): 1203–7.

Wilson, Edmund O. *The Diversity of Life.* New York: W. W. Norton, 1992.

—*The Creation.* New York: W. W. Norton, 2006.

World Council of Churches. *Signs of the Spirit* (1990 Canberra Assembly), Michael Kinnamon, ed. Grand Rapids: Eerdmans, 1991.

Życiński, Józef. *God and Evolution.* Washington, DC: Catholic University of America Press, 2006.

Zygon: Journal of Religion and Science 46/2 (June 2011): "Responses to Darwin in the Religious Traditions," 265–395.

INDEX